Evolution: A Very Short Introduction

VERY SHORT INTRODUCTIONS are for anyone wanting a stimulating and accessible way in to a new subject. They are written by experts, and have been published in more than 25 languages worldwide.

The series began in 1995, and now represents a wide variety of topics in history, philosophy, religion, science, and the humanities. Over the next few years it will grow to a library of around 200 volumes – a Very Short Introduction to everything from ancient Egypt and Indian philosophy to conceptual art and cosmology.

Very Short Introductions available now:

ANCIENT PHILOSOPHY Julia Annas
THE ANGLO-SAXON AGE John Blair
ANIMAL RIGHTS David DeGrazia
ARCHAEOLOGY Paul Bahn
ARCHITECTURE Andrew Ballantyne
ARISTOTLE Jonathan Barnes
ART THEORY Cynthia Freeland
THE HISTORY OF ASTRONOMY Michael Hoskin
ATHEISM Julian Baggini
AUGUSTINE Henry Chadwick
BARTHES Jonathan Culler
THE BIBLE John Riches
BRITISH POLITICS Anthony Wright
BUDDHA Michael Carrithers
BUDDHISM Damien Keown
THE CELTS Barry Cunliffe
CHOICE THEORY Michael Allingham
CLASSICS Mary Beard and John Henderson
CLAUSEWITZ Michael Howard
THE COLD WAR Robert McMahon
CONTINENTAL PHILOSOPHY Simon Critchley
COSMOLOGY Peter Coles
CRYPTOGRAPHY Fred Piper and Sean Murphy
DARWIN Jonathan Howard
DEMOCRACY Bernard Crick
DESCARTES Tom Sorell
DRUGS Leslie Iversen
THE EARTH Martin Redfern
EIGHTEENTH-CENTURY BRITAIN Paul Langford
EMOTION Dylan Evans
EMPIRE Stephen Howe
ENGELS Terrell Carver
ETHICS Simon Blackburn
THE EUROPEAN UNION John Pinder
EVOLUTION Brian and Deborah Charlesworth
FASCISM Kevin Passmore
THE FRENCH REVOLUTION William Doyle
FREUD Anthony Storr
GALILEO Stillman Drake
GANDHI Bhikhu Parekh
GLOBALIZATION Manfred Steger
HEGEL Peter Singer
HEIDEGGER Michael Inwood
HINDUISM Kim Knott

HISTORY John H. Arnold
HOBBES Richard Tuck
HUME A. J. Ayer
IDEOLOGY Michael Freeden
INDIAN PHILOSOPHY Sue Hamilton
INTELLIGENCE Ian J. Deary
ISLAM Malise Ruthven
JUDAISM Norman Solomon
JUNG Anthony Stevens
KANT Roger Scruton
KIERKEGAARD Patrick Gardiner
THE KORAN Michael Cook
LINGUISTICS Peter Matthews
LITERARY THEORY Jonathan Culler
LOCKE John Dunn
LOGIC Graham Priest
MACHIAVELLI Quentin Skinner
MARX Peter Singer
MATHEMATICS Timothy Gowers
MEDIEVAL BRITAIN John Gillingham and Ralph A. Griffiths
MODERN IRELAND Senia Pašeta
MUSIC Nicholas Cook
NIETZSCHE Michael Tanner
NINETEENTH-CENTURY BRITAIN Christopher Harvie and H. C. G. Matthew
NORTHERN IRELAND Marc Mulholland
PAUL E. P. Sanders
PHILOSOPHY Edward Craig
PHILOSOPHY OF SCIENCE Samir Okasha
PLATO Julia Annas
POLITICS Kenneth Minogue
POLITICAL PHILOSOPHY David Miller
POSTCOLONIALISM Robert Young
POSTMODERNISM Christopher Butler
POSTSTRUCTURALISM Catherine Belsey
PREHISTORY Chris Gosden
PSYCHOLOGY Gillian Butler and Freda McManus
QUANTUM THEORY John Polkinghorne
ROMAN BRITAIN Peter Salway
ROUSSEAU Robert Wokler
RUSSELL A. C. Grayling
RUSSIAN LITERATURE Catriona Kelly
THE RUSSIAN REVOLUTION S. A. Smith
SCHIZOPHRENIA Chris Frith and Eve Johnstone
SCHOPENHAUER Christopher Janaway
SHAKESPEARE Germaine Greer
SOCIAL AND CULTURAL ANTHROPOLOGY John Monaghan and Peter Just
SOCIOLOGY Steve Bruce
SOCRATES C. C. W. Taylor
SPINOZA Roger Scruton
STUART BRITAIN John Morrill
TERRORISM Charles Townshend
THEOLOGY David F. Ford
THE TUDORS John Guy
TWENTIETH-CENTURY BRITAIN Kenneth O. Morgan
WITTGENSTEIN A. C. Grayling
WORLD MUSIC Philip Bohlman

Available soon:

AFRICAN HISTORY John Parker and Richard Rathbone
ANCIENT EGYPT Ian Shaw
ART HISTORY Dana Arnold
THE BRAIN Michael O'Shea
BUDDHIST ETHICS Damien Keown
CAPITALISM James Fulcher
CHAOS Leonard Smith
CHRISTIAN ART Beth Williamson
CHRISTIANITY Linda Woodhead
CITIZENSHIP Richard Bellamy
CLASSICAL ARCHITECTURE Robert Tavernor
CLONING Arlene Judith Klotzko
CONTEMPORARY ART Julian Stallabrass
THE CRUSADES Christopher Tyerman
DADA AND SURREALISM David Hopkins
DERRIDA Simon Glendinning
DESIGN John Heskett
DINOSAURS David Norman
DREAMING J. Allan Hobson
ECONOMICS Partha Dasgupta
EGYPTIAN MYTHOLOGY Geraldine Pinch
THE ELEMENTS Philip Ball
THE END OF THE WORLD Bill McGuire
EXISTENTIALISM Thomas Flynn
THE FIRST WORLD WAR Michael Howard
FREE WILL Thomas Pink
FUNDAMENTALISM Malise Ruthven
HABERMAS Gordon Finlayson
HIEROGLYPHS Penelope Wilson
HIROSHIMA B. R. Tomlinson
HUMAN EVOLUTION Bernard Wood
INTERNATIONAL RELATIONS Paul Wilkinson
JAZZ Brian Morton
MANDELA Tom Lodge
MEDICAL ETHICS Tony Hope
THE MIND Martin Davies
MOLECULES Philip Ball
MYTH Robert Segal
NATIONALISM Steven Grosby
PERCEPTION Richard Gregory
PHILOSOPHY OF RELIGION Jack Copeland and Diane Proudfoot
PHOTOGRAPHY Steve Edwards
THE PRESOCRATICS Catherine Osborne
THE RAJ Denis Judd
THE RENAISSANCE Jerry Brotton
RENAISSANCE ART Geraldine Johnson
SARTRE Christina Howells
THE SPANISH CIVIL WAR Helen Graham
TRAGEDY Adrian Poole
THE TWENTIETH CENTURY Martin Conway

For more information visit our web site
www.oup.co.uk/vsi

Brian and Deborah Charlesworth

EVOLUTION

A Very Short Introduction

OXFORD
UNIVERSITY PRESS

Great Clarendon Street, Oxford OX2 6DP

Oxford University Press is a department of the University of Oxford.
It furthers the University's objective of excellence in research, scholarship,
and education by publishing worldwide in

Oxford New York

Auckland Bangkok Buenos Aires Cape Town Chennai
Dar es Salaam Delhi Hong Kong Istanbul Karachi Kolkata
Kuala Lumpur Madrid Melbourne Mexico City Mumbai Nairobi
São Paulo Shanghai Taipei Tokyo Toronto

Published in the United States
by Oxford University Press Inc., New York

First published as a Very Short Introduction 2003

British Library Cataloguing in Publication Data

Data available

Library of Congress Cataloging in Publication Data

Data available

ISBN 0-19-280251-8

1 3 5 7 9 10 8 6 4 2

Typeset by RefineCatch Ltd, Bungay, Suffolk
Printed in Spain by Book Print S. L., Barcelona

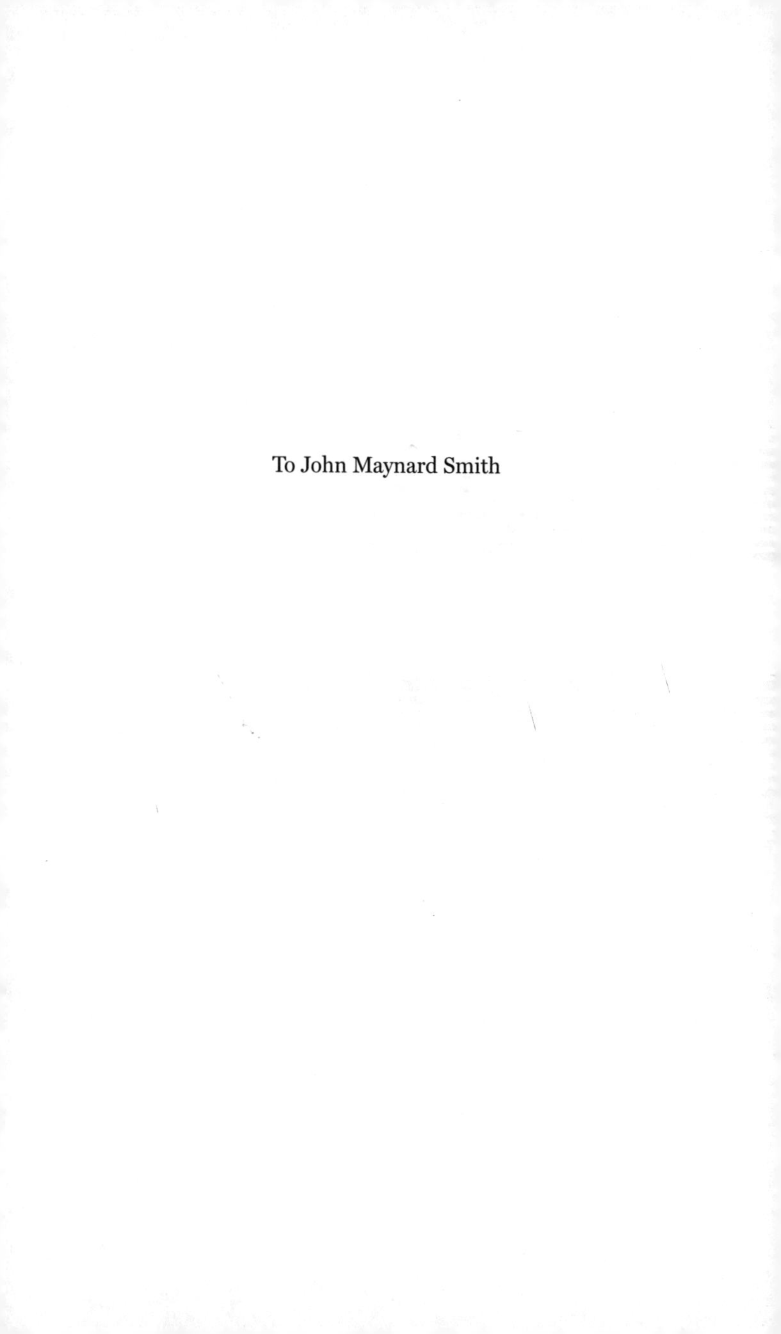

To John Maynard Smith

Acknowledgements

We thank Shelley Cox and Emma Simmons of Oxford University Press for respectively suggesting that we write this book and for editing it. We also thank Helen Borthwick, Jane Charlesworth, and John Maynard Smith for reading and commenting on the first draft of the manuscript. All remaining errors are, of course, our fault.

Contents

List of illustrations

The publisher and the author apologize for any errors or omissions in the above list. If contacted they will be pleased to rectify these at the earliest opportunity.

Chapter 1
Introduction

> We are all one with creeping things;
> And apes and men
> Blood-brethren.
>
> From 'Drinking Song' by Thomas Hardy

The consensus among the scientific community is that the Earth is a planet orbiting a fairly typical star, one of many billions of stars in a galaxy among billions of galaxies in an expanding universe of enormous size, which originated about 14 billion years ago. The Earth itself formed as the result of a process of gravitational condensation of dust and gas, which also generated the Sun and other planets of the solar system, about 4.6 billion years ago. All present-day living organisms are the descendants of self-replicating molecules that were formed by purely chemical means, more than 3.5 billion years ago. The successive forms of life have been produced by the process of 'descent with modification', as Darwin called it, and are related to each other by a branching genealogy, the tree of life. We human beings are most closely related to chimpanzees and gorillas, with whom we shared a common ancestor 6 to 7 million years ago. The mammals, the group to which we belong, shared a common ancestor with living species of reptiles about 300 million years ago. All vertebrates (mammals, birds, reptiles, amphibia, fishes) trace their ancestry back to a small fish-like creature that lacked a backbone, which lived over 500 million years ago. Further back in time, it becomes increasingly difficult to discern the relationships between the major groups of animals, plants, and microbes, but, as we shall see, there are clear signs in their genetic material of common ancestry.

Less than 450 years ago, all European scholars believed that the Earth was the centre of a universe of at most a few million miles in extent, and that the planets, Sun, and stars all rotated around this centre. Less than 250 years ago, they believed that the universe was created in essentially its present state about 6,000 years ago, although by then the Earth was known to orbit the Sun like other planets, and a much larger size of the universe was widely accepted. Less than 150 years ago, the view that the present state of the Earth is the product of at least tens of millions of years of geological change was prevalent among scientists, but the special creation by God of living species was still the dominant belief.

The relentless application of the scientific method of inference from experiment and observation, without reference to religious or governmental authority, has completely transformed our view of our origins and relation to the universe, in less than 500 years. In addition to the intrinsic fascination of the view of the world opened up by science, this has had an enormous impact on philosophy and religion. The findings of science imply that human beings are the product of impersonal forces, and that the habitable world forms a minute part of a universe of immense size and duration. Whatever the religious or philosophical beliefs of individual scientists, the whole programme of scientific research is founded on the assumption that the universe can be understood on such a basis.

Few would dispute that this programme has been spectacularly successful, particularly in the 20th century, which saw such terrible events in human affairs. The influence of science may have indirectly contributed to these events, partly through the social changes triggered by the rise of industrial mass societies, and partly through the undermining of traditional belief systems. Nonetheless, it can be argued that much misery throughout human history could have been avoided by the application of reason, and that the disasters of the 20th century resulted from a failure to be rational rather than a failure of rationality. The wise application of scientific understanding of the world in which we live is the only hope for the future of mankind.

The study of evolution has revealed our intimate connections with the other species that inhabit the Earth; if global catastrophe is to be avoided, these connections must be respected. The purpose of this book is to introduce the general reader to some of the most important basic findings, concepts, and procedures of evolutionary biology, as it has developed since the first publications of Darwin and Wallace on the subject, over 140 years ago. Evolution provides a set of unifying principles for the whole of biology; it also illuminates the relation of human beings to the universe and to each other. In addition, many aspects of evolution have practical importance; for instance, pressing medical problems are posed by the rapid evolution of resistance by bacteria to antibiotics and of HIV to antiviral drugs.

In this book, we shall first introduce the main causal processes of evolution (Chapter 2). Chapter 3 provides some of the basic biological background, and shows how the similarities between living creatures can be understood in terms of evolution. Chapter 4 describes the evidence for evolution derived from Earth history, and from the patterns of geographical distribution of living species. Chapter 5 is concerned with the evolution of adaptations by natural selection, and Chapter 6 with the evolution of new species and of differences between species. In Chapter 7, we discuss some seemingly difficult problems for the theory of evolution. Chapter 8 provides a brief summary.

Chapter 2
The processes of evolution

To understand life on Earth, we need to know how animals (including humans), plants, and microbes work, ultimately in terms of the molecular processes that underlie their functioning. This is the 'how' question of biology; an enormous amount of research during the last century has produced spectacular progress towards answering this question. This effort has shown that even the simplest organism capable of independent existence, a bacterial cell, is a machine of great complexity, with thousands of different protein molecules that act in a coordinated fashion to fulfil the functions necessary for the cell to survive, and to divide to produce two daughter cells (see Chapter 3). This complexity is even greater in higher organisms such as a fly or human being. These start life as a single cell, formed by the fusion of an egg and a sperm. There is then a delicately controlled series of cell divisions, accompanied by the differentiation of the resulting cells into many distinct types. The process of development eventually produces the adult organism, with its highly organized structure made up of different tissues and organs, and its capacity for elaborate behaviour. Our understanding of the molecular mechanisms that underlie this complexity of structure and function is rapidly expanding. Although there are still many unsolved problems, biologists are convinced that even the most complicated features of living creatures, such as human consciousness, reflect the operation of chemical and physical processes that are accessible to scientific analysis.

At all levels, from the structure and function of a single protein molecule, to the organization of the human brain, we see many instances of *adaptation*: the fit of structure to function that is also apparent in machines designed by people (see Chapter 5). We also see that different species have distinctive characteristics, often clearly reflecting adaptations to the environments in which they live. These observations raise the 'why' question of biology, which concerns the processes that have caused organisms to be the way they are. Before the rise of the idea of evolution, most biologists would have answered this question by appealing to a Creator. The term adaptation was introduced by 18th-century British theologians, who argued that the appearance of design in the features of living creatures proves the existence of a supernatural designer. While this argument was shown to be logically flawed by the philosopher David Hume in the middle of the 18th century, it retained its hold on people's minds as long as no credible alternative had been proposed.

Evolutionary ideas provide a set of natural processes that can explain the vast diversity of living species, and the characteristics that make them so well adapted to their environment, without any appeal to supernatural intervention. These explanations extend, of course, to the origin of the human species itself, and this has made biological evolution the most controversial of scientific subjects. If the issues are approached without prejudice, however, the evidence for evolution as an historical process can be seen to be as strong as that for other long-established scientific theories, such as the atomic nature of matter (see Chapters 3 and 4). We also have a set of well-verified ideas about the causes of evolution, although, as in every healthy science, there are unsolved problems, as well as new questions that arise as more is understood (see Chapter 7).

Biological evolution involves changes over time in the characteristics of populations of living organisms. The time-scale and magnitude of such changes vary enormously. Evolution can be studied during a human lifetime, when simple changes occur in a single character, such as the increase in the frequency of strains of bacteria resistant to penicillin within a few years of the widespread medical use of penicillin to control bacterial infections (as discussed in Chapter 5). At the other extreme, evolution involves events such as the emergence of a major new design of organisms, which may take millions of years and require changes in many different characteristics, as in the transition from reptiles to mammals (see Chapter 4). A key insight of the founders of evolutionary theory, Charles Darwin and Alfred Russel Wallace, was that changes at all levels are likely to involve the same types of processes. Major evolutionary changes largely reflect changes of the same type as more minor events, accumulated over longer time periods (see Chapters 6 and 7).

Evolutionary change ultimately relies on the appearance of new variant forms of organisms: *mutations*. These are caused by stable changes in the genetic material, transmitted from parent to offspring. Mutations affecting almost all conceivable characteristics of many different organisms have been studied in the laboratory by experimental geneticists, and medical geneticists have catalogued thousands of mutations in human populations. The effects of mutations on the observable characteristics of an organism vary greatly in their magnitude. Some have no detectable effect, and are known to exist simply because it is now possible to study the structure of the genetic material directly, as we will describe in Chapter 3. Others have relatively small effects on a simple trait, such as a change in eye colour from brown to blue, the acquisition of resistance to an antibiotic by a bacterium, or an alteration of the number of bristles on the side of a fruitfly. Some mutations have drastic effects on development, such as the mutation of the fruitfly *Drosophila melanogaster* that causes a leg to grow on the fly's head in place of its antenna. The appearance of any particular kind of new mutation is a very rare event, with a frequency of around one per hundred thousand individuals per generation or even less. An altered state of a character as a result of a mutation, such as antibiotic resistance, initially occurs in a single individual, and is usually restricted to a tiny fraction of a typical population for many generations. To result in evolutionary change, other processes must cause it to increase in frequency within the population.

Natural selection is the most important of these processes for evolutionary changes that involve the structure, functioning, and behaviour of organisms (see Chapter 5). In their papers of 1858, published in the *Journal of the Proceedings of the Linnaean Society*, Darwin and Wallace laid out their theory of evolution by natural selection with the following argument:

- Many more individuals of a species are born than can normally live to maturity and breed successfully, so that there is a *struggle for existence.*
- There is *individual variation* in innumerable characteristics of the population, some of which may affect an individual's ability to survive and reproduce. The successful parents of a given generation may therefore differ from the population as a whole.
- There is likely to be a *hereditary component* to much of this variation, so that the characteristics of the offspring of the successful parents will differ from the characteristics of the previous generation, in a similar way to their parents.

If this process continues from generation to generation, there will be a gradual transformation of the population, such that the frequencies of characteristics associated with greater survival ability or reproductive success increase over time. These altered characteristics originated by mutation, but mutations affecting a particular trait arise all the time regardless of whether or not they are favoured by selection. Indeed, most mutations either have no effects on the organism, or reduce its ability to survive or reproduce.

It is the process of increase in frequency of variants that improve survival or reproductive success that explains the evolution of adaptive characteristics, since better performance of the individual's body or behaviour will generally contribute to greater survival or reproductive success. Such a process of change will be especially likely if a population is exposed to a changed environment, where a somewhat different set of characteristics is favoured from those already established by selection. As Darwin wrote in 1858:

> But let the external conditions of a country alter . . . Now, can it be doubted, from the struggle each individual has to obtain subsistence, that any minute variation in structure, habits or instincts, adapting that individual better to the new conditions, would tell upon its vigour and health? In the struggle it would have a better *chance* of surviving; and those of its offspring that inherited the variation, be it ever so slight, would also have a better *chance*. Yearly more are bred than can survive; the smallest grain in the balance, in the long run, must tell on which death shall fall, and which shall survive. Let this work of selection on the one hand, and death on the other, go on for a thousand generations, who will pretend to affirm that it would produce no effect . . .

There is, however, another important mechanism of evolutionary change, which explains how species can also come to differ with respect to traits with little or no influence on the survival or reproductive success of their possessors, and which are therefore not subject to natural selection. As we shall see in Chapter 6, this is especially likely to be true of the large category of changes in the genetic material which have little or no effect on the organism's structure or functioning. If there is *selectively neutral* variability, so that on average there are no differences in survival or fertility among different individuals, it is still possible for the offspring generation to differ slightly from the parental generation. This is because, in the absence of selection, the genes in the population of offspring are a random sample of the genes present in the parental population. Real populations are finite in size, and so the constitution of the offspring population will by chance differ somewhat from that of the parents' generation, just as we do not expect exactly five heads and five tails when we toss a coin ten times.

This process of random change is called *genetic drift*. Even the biggest biological populations, such as those of bacteria, are finite, so that genetic drift will always operate.

The combined effects of mutation, natural selection and the random process of genetic drift cause changes in the composition of a population. Over a sufficiently long period of time, these cumulative effects alter the population's genetic make-up, and can thus greatly change the species' characteristics from those of its ancestors.

We referred earlier to the diversity of life, reflected in the large number of different species alive today. (A very much larger number have existed over the past history of life, owing to the fact that the ultimate fate of nearly all species is extinction, as described in Chapter 4.) The problem of how new species evolve is clearly a crucial one, and is dealt with in Chapter 6. The term 'species' is hard to define, and it is sometimes difficult to draw a clear line between populations that are members of the same species, and populations that belong to separate species. In thinking about evolution, it makes sense to consider two populations of sexually reproducing organisms as different species if they cannot interbreed with each other, so that their evolutionary fates are totally independent. Thus, human populations living in different parts of the world are unequivocally members of the same species, since there are no barriers to interbreeding if migrant individuals arrive from another place. Such migration tends to prevent the genetic makeup of different populations of the same species from diverging very much. In contrast, chimpanzees and humans are clearly separate species, since humans and chimpanzees living in the same area cannot interbreed. As we shall describe later on, humans also differ much more from chimpanzees in the make-up of their genetic material than they do from each other. The formation of a new species must involve the evolution of barriers to interbreeding between related populations. Once such barriers form, the populations can diverge under mutation, selection, and genetic drift. This process of divergence ultimately leads to the diversity of life. If we understand how barriers to interbreeding evolve, and how populations subsequently diverge, we will understand the origin of species.

An enormous amount of biological data falls into place in the light of these ideas about evolution, which have been put on a firm basis by the development of mathematical theories which can be modelled in detail, just as astronomers and physicists model the behaviour of stars, planets, molecules, and atoms in order to understand them more completely, and to devise detailed tests of their theories. Before describing the mechanisms of evolution in more detail (but omitting the mathematics), the next two chapters will show how many kinds of biological observations make sense in terms of evolution, in contrast with special creation and its appeal to *ad hoc* explanations.

Chapter 3
The evidence for evolution: similarities and differences between organisms

The theory of evolution accounts for the diversity of life, with all the well-known differences between different species of animals, plants, and microbes, but it also explains their fundamental similarities. These are often evident at the superficial level of externally visible characters, but extend to the finest details of microscopic structure and biochemical function. We will discuss the diversity of life later in this book (in Chapter 6), and describe how the theory of evolution can account for new forms appearing from ancestral ones, but here we focus on the unity of living species. In addition, we will introduce many basic biological facts on which later chapters build.

Similarities between different groups of species

Similarities between even widely disparate types of organism exist at every level, from familiar, externally visible resemblances, to profound resemblances in life-cycles and the structure of the genetic material. They are plainly detectable even between creatures as different as ourselves and bacteria. These similarities have a natural and straightforward explanation in the idea that organisms are related through an evolutionary process of descent from common ancestors. We ourselves have obvious similarities to apes, as illustrated in Figure 1A, including similarities in internal characters such as our brain structure and organization. There are

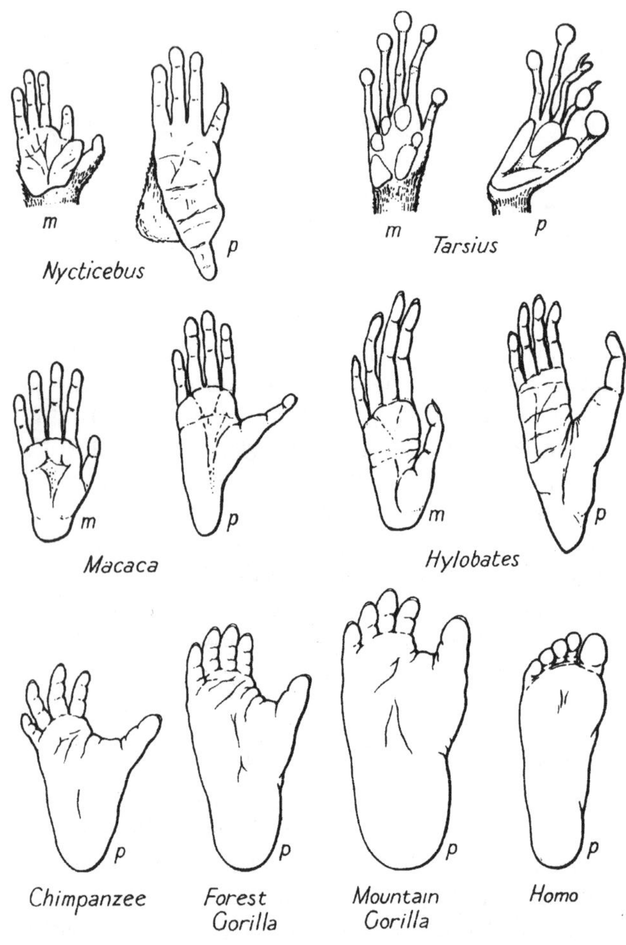

1. A. Hands (*m*) and feet (*p*) of several primate species, showing the similarities between different species, with differences related to the animals' way of life, such as the opposable digits of climbing species (*Hylobates* is a gibbon, *Macaca* is a Rhesus monkey, *Nycticebus* and *Tarsius* are primitive arboreal primates). B. Skeletons of a bat and a bird, showing their similarities and differences.

B

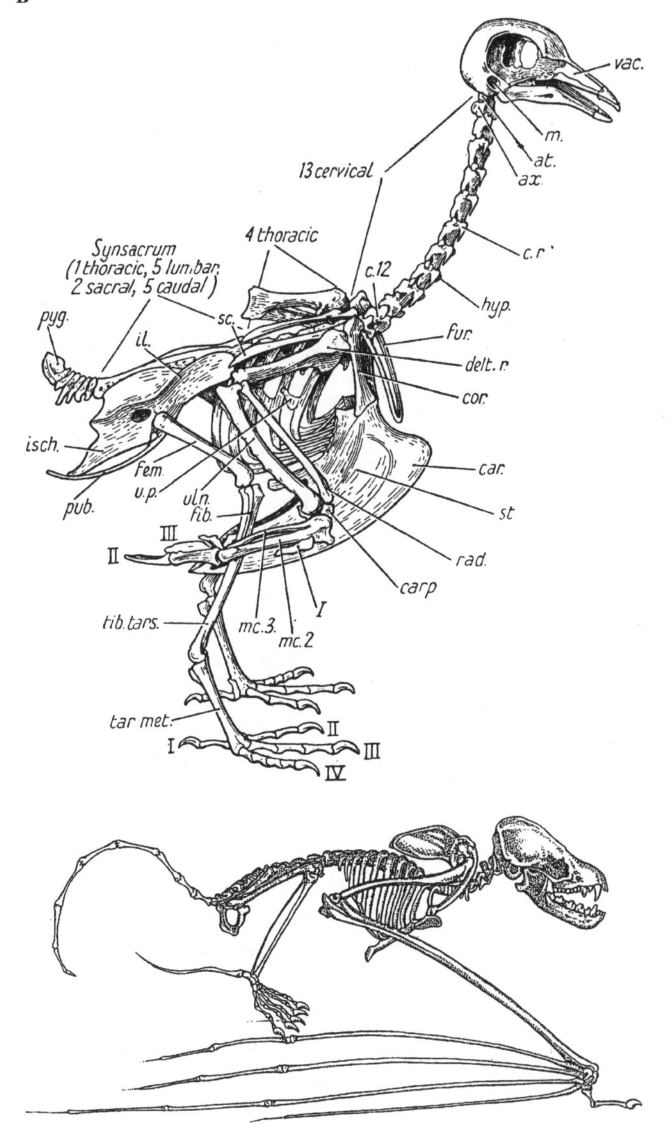

vac.
m.
at.
ax.
13 cervical
c.r.
4 thoracic
c.12
hyp.
Synsacrum
(1 thoracic, 5 lumbar,
2 sacral, 5 caudal)
pyg.
sc.
il.
fur.
delt. r
cor.
isch.
fem.
car.
u.p.
uln.
st
pub.
fib.
III
II
rad.
carp
I
tib. tars.
mc.3.
mc.2
tar met.
II
I
III
IV

lesser similarities to monkeys, and even smaller, but still extremely clear, similarities to other mammals, despite all our differences. Mammals have many similarities to other vertebrates, including the basic features of their skeletons, and their digestive, circulatory, and nervous systems. Even more amazing are similarities with creatures such as insects, for example in their segmented body plans, their common need for sleep, the control of their daily rhythms of sleep and waking, and fundamental similarities in how the nerves work in many different kinds of animals, among other features.

Systems of biological classification have long been based on easily visible structural characteristics. For example, even before the scientific study of biology, insects were treated as a group of similar creatures, clearly distinguishable from other groups of invertebrates, such as molluscs, by their possession of a segmented body, six pairs of jointed legs, a tough external protective covering, and so on. Many of these traits are shared with other types of animal such as crabs and spiders, except that the numbers of legs may differ (eight, in the case of spiders). These different species are all grouped into one larger division, the arthropods. The arthropods include the insects, and among these flies form one group, characterized by the fact that they all have only one pair of wings, as well as several other shared characters. Butterflies and moths form another insect group, whose members all have fine scales on their two pairs of wings. Among flies we distinguish the houseflies and their relatives from other groups by shared characters, and among these we name individual *species*, such as the common housefly *Musca domestica*. Species are essentially groups of similar individuals capable of interbreeding with each other. Similar species are grouped into the same *genus*, again united by a set of characters not shared with other genera. Biologists identify each recognizable species by two names, the genus name followed by the name of the species itself, for example *Homo sapiens*; these names are conventionally written in italics.

The observation that organisms can be classified hierarchically into groups, which successively share more and more traits that are lacking in other groups, was an important advance in biology. The classification of organisms into species, and the naming system for species, were developed long before Darwin. Before biologists could begin thinking about the evolution of species, it was clearly important to have the concept of species as distinct entities. The simplest and most natural way to account for the hierarchical pattern of similarities is that living organisms evolved over time, starting from ancestral forms that diversified to produce the groups alive today, as well as innumerable extinct organisms (see Chapter 4). As we shall discuss in Chapter 6, it is now possible to discern this inferred pattern of genealogical relationships among groups of organisms by directly studying the information in their genetic material.

Another set of facts that strongly supports the theory of evolution is provided by modifications of the same structure in different species. For instance, the bones of bats' and birds' wings indicate clearly that they are modified forelimbs, even though they look very different from the forelimbs of other vertebrates (Figure 1B). Similarly, although the flippers of whales look much like fish fins, and are clearly also well adapted for swimming, their internal structure is like the feet of other mammals, except for an increased number of digits. This makes sense, given all the other evidence that whales are modified mammals (for instance, they breathe with lungs and suckle their young). Fossil evidence shows that the two pairs of limbs of land vertebrates are derived from the two pairs of fins of the lobe-finned fishes (of which coelacanths are the most famous living representatives, see Chapter 4). Indeed, the earliest land vertebrate fossils had more than five digits on their limbs, just like fishes and whales. Another example is provided by the three small bones in mammals' ears, which transmit sound from the outside to the organ that transforms sound into nerve signals. These tiny bones develop from rudiments in the embryonic jaw and skull, and in reptiles they enlarge during development to make parts of the head and jaw skeleton. Fossil intermediates that connect reptiles with mammals show successive modifications of these bones in the adults, finally evolving into the ear bones. These examples are just a few of many known cases in which the same basic structure was considerably modified during the course of evolution by the demands imposed by different functions.

Embryonic development and vestigial organs

Embryonic development provides many other striking examples of similarities between different groups of organisms, clearly suggesting descent from common ancestors. The embryonic forms of different species are often extremely similar, even when the adults are very different. For example, at one stage in mammalian development, gill slits appear that resemble those of fish embryos (Figure 2). This makes perfect sense if we are descended from fish-like ancestors, but is otherwise inexplicable. Since it is the adult structures that adapt the organism to its environment, they are very likely to be modified by selection. Probably the developing blood vessels require the presence of gill slits to guide them to form in the correct places, so that these structures are retained, even in animals that never have functional gills. Development can evolve, however. In many other details, mammals develop very differently from fish, so that other embryonic structures, with less profound importance in development, have been lost, and new ones have been gained.

Similarities are not confined to embryonic stages. *Vestigial organs* have also long been recognized as remnants of structures that were functional in the ancestors of present-day organisms. Their evolution is very interesting, because such cases tell us that evolution does not always create and improve structures, but sometimes reduces them. The human appendix, which is a greatly reduced version of a part of the digestive tract that is quite large in

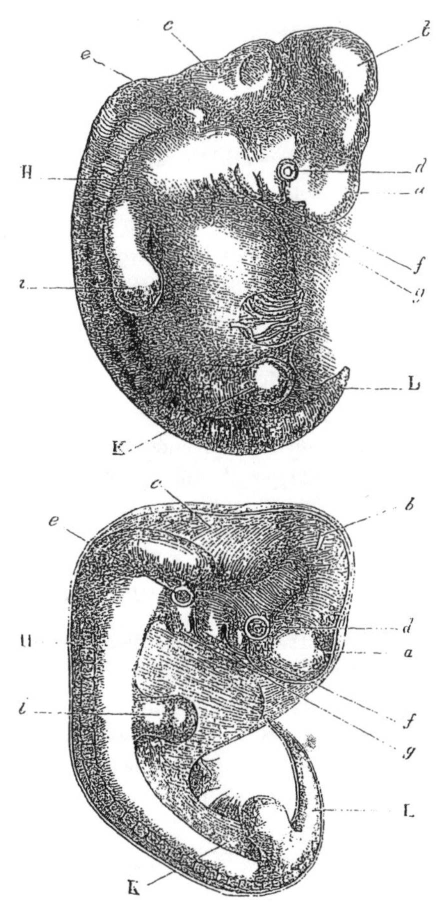

Fig. 1. Upper figure human embryo, from Ecker. Lower figure that of a dog, from Bischoff.

a. Fore-brain, cerebral hemispheres, &c.
b. Mid-brain, corpora quadrigemina.
c. Hind-brain, cerebellum, medulla oblongata.
d. Eye.
e. Ear.
f. First visceral arch.
g. Second visceral arch.
H. Vertebral columns and muscles in process of development.
i. Anterior } extremities.
K. Posterior } extremities.
L. Tail or os coccyx.

2. Human and dog embryos, illustrating their great similarity at this stage of development. The gill slits, labelled visceral arches (f and g) in the figure, are plainly visible. From Darwin's *The Descent of Man and Selection in Relation to Sex* (1871).

orang-utans, is a classic example. The vestigial limbs of legless animals are also well known. Fossils of primitive snakes have been found with almost complete hindlimbs, indicating that snakes evolved from lizard-like ancestors with legs. The body of a present-day snake consists of an elongated thorax (chest), with a large number of vertebrae (more than 300 in pythons). In the python, the change from the body to tail is marked by vertebrae with no ribs, and at this point rudimentary hindlimbs are found. There is a pelvic girdle and a pair of truncated thigh bones whose development follows the normal course for other vertebrates, with expression of the same genes that normally control limb development. A graft of python hindlimb tissue can even promote the formation of an extra digit in chick wings, showing that parts of the hindlimb developmental system still exist in pythons. More advanced types of snakes, however, are completely limbless.

Similarities in cells and cellular functions

The similarities between different organisms are not confined to visible characteristics. They are profound and extend to the smallest microscopic scale and to the most fundamental aspects of life. A basic feature of all animal, plant, and fungal life is that their tissues are made up of essentially similar units, the *cells*. Cells are the basis of the bodies of all organisms other than viruses, from unicellular yeasts and bacteria, to multicellular bodies with highly differentiated tissues like those of mammals. In the *eukaryotes* (all cellular non-bacterial life) the cells are organized into the *cytoplasm* and the *nucleus* within it that contains the genetic material (Figure 3). The cytoplasm is not just a liquid inside the cell membrane with the nucleus floating in it; it contains a complex set of tiny pieces of machinery that includes many subcellular structures. Two of the most important of these cellular *organelles* are the mitochondria that generate cells' energy, and the chloroplasts in which photosynthesis in green plants' cells occurs. It is now known that both these are descended from bacteria that colonized cells and became integrated into them as essential components. Bacteria are also cells (Figure 3), but simpler ones with no nucleus or organelles; they and similar organisms are called *prokaryotes*. The only non-cellular forms of life, the viruses, are parasites that reproduce inside the cells of other organisms, and consist simply of a protein coat surrounding the genetic material.

Cells are ultra-miniaturized and highly complex factories which make the chemicals that organisms need, generate energy from food sources, and produce bodily structures such as the bones of animals. Most of the 'machines' and many of the structures in these factories are *proteins*. Some proteins are *enzymes* that take a chemical and carry out a procedure on it, for example snipping a chemical compound into two components, like chemical scissors. The enzymes used in biological detergents snip up proteins (such as blood and sweat proteins) into small pieces that can be washed out of dirty clothes; similar enzymes in our gut break molecules in food into smaller pieces that can be taken up by cells. Other proteins in living organisms have storage or transport functions. The haemoglobin in red blood cells carries oxygen, and in the liver a protein called ferritin binds and stores iron. There are also structural proteins, such as the keratin that forms skin, hair, and fingernails. In addition, cells make proteins that communicate information to other cells and to other organs. Hormones are familiar communication proteins, which circulate in the blood and control many bodily functions. Other proteins are located on cell surfaces and are involved in communication with other cells. These interactions include signalling to control cell behaviour during development, communication between eggs and sperm in fertilization, and parasite recognition by the immune system.

Like any factory, cells are subject to complex controls. They respond to information from outside (by means of proteins that span the cell membrane, like keyholes which fit molecules from the outside

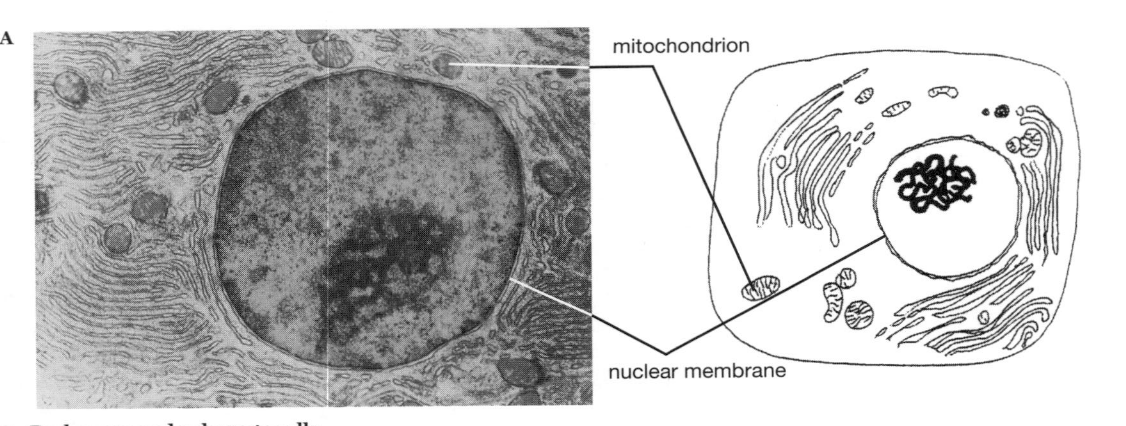

3. Prokaryote and eukaryote cells.

A. Electron micrograph and drawing of a portion of a cell from the mammalian pancreas, showing the nucleus containing the chromosomes inside the nuclear membrane, the region outside the nucleus containing many mitochondria (these organelles also have membranes enclosing them), and membrane-like structures that are involved in protein synthesis and export, as well as in importing substances into the cell. A mitochondrion is somewhat smaller than a bacterial cell.

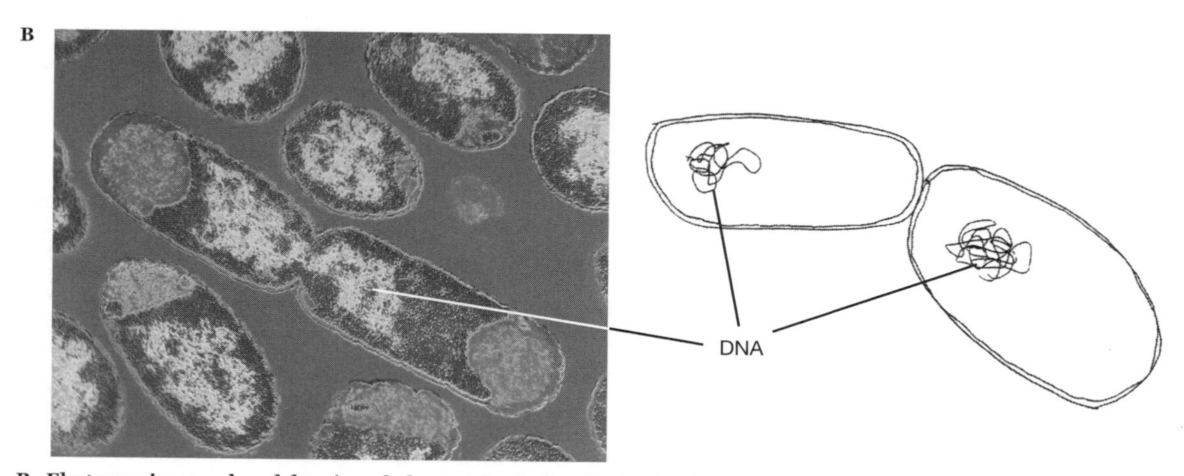

B. Electron micrograph and drawing of a bacterial cell, showing its simple structure, with a cell wall and DNA which is not enclosed in a nucleus.

world – see Figure 4). Sensory receptor proteins, such as olfactory receptors and light receptors, are used in communication between cells and their environment. Chemical and light signals from the outside world are transformed into electrical impulses that travel along the nerves to the brain. All animals that have been studied use largely similar proteins in chemical and light perception. To illustrate the similarities that have been discovered in cells of different organisms, a myosin (motor) protein, similar to proteins in muscle cells, is involved in signalling in flies' eyes and in the ears of humans; one form of deafness is caused by mutations in the gene for this protein.

Biochemists have catalogued the enzymes in living organisms into many different kinds, and every known enzyme (many thousands in a complex animal like ourselves) has a number in an international numbering system. Because so many enzymes are found in cells of a very wide range of organisms, this system categorizes enzymes by the jobs they perform, not the organism they come from. Some, such as digestive enzymes, snip molecules into pieces, others combine molecules together, while others oxidize chemicals (combine them with oxygen), and so on.

The means by which energy is generated by cells from food sources is largely the same for all kinds of cells. In this process, there is an energy source (sugars or fats, in the case of our cells, but other compounds, such as hydrogen sulphide, for some bacteria). A cell takes the initial compound through a series of chemical steps, some of which release energy. Such a *metabolic pathway* is organized like an assembly line, with a succession of sub-processes. Each sub-process is carried out by its own protein 'machine'; these are the enzymes for the different steps in the pathway. The same pathways operate in a wide range of organisms, and modern biology textbooks show the important metabolic pathways without needing to specify the organism. For example, when lizards tire after running, this is caused by the build-up of the chemical lactic acid, just as in our muscles. Cells have pathways to make chemicals of many different kinds, as well as to generate energy from foods. For example, some of our cells make hairs, some make bone, some make pigments, others produce hormones, and so on. The metabolic pathway by which the skin pigment melanin is made (Figure 4) is the same in ourselves, in other mammals, in butterflies with black wing pigments, and even in fungi (for instance in black spores), and many of the enzymes involved in this pathway are also used by plants in making lignin, the main chemical constituent of wood. The fundamental similarity of the basic features of metabolic pathways, from bacteria to mammals, is once again readily understandable in terms of evolution.

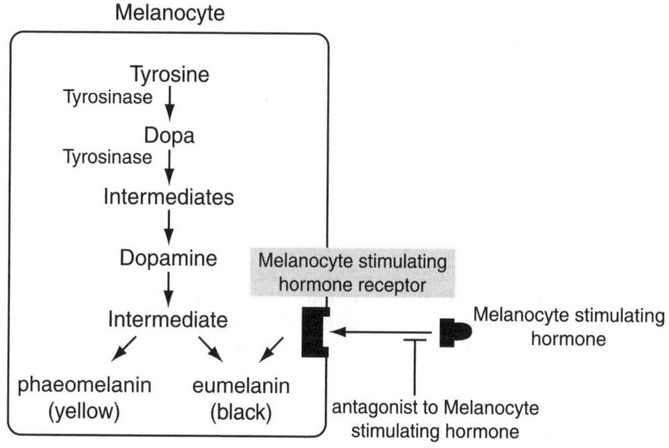

4. Biosynthetic pathways by which melanin and a yellow pigment are synthesized in mammalian melanocyte cells from their amino acid precursor, tyrosine. Each step in the pathway is catalysed by a different enzyme. Absence of active tyrosinase enzyme results in albino animals. The melanocyte-stimulating hormone receptor determines the relative amounts of black and yellow pigments. Absence of the antagonist to the hormone leads to black pigment synthesis, but presence of the antagonist sets the receptor to 'off', leading to yellow pigment formation. This is how the yellow versus black parts of tabby cat and brown mouse hairs come to be formed. Mutations that make the antagonist non-functional cause darker coloration; however, black animals are not usually the result of this, but simply have the receptor set to 'on' regardless of the hormone level.

Each of the different proteins for these cell and body functions is specified by one of the organism's genes, as we will explain more fully later in this chapter. The functioning of each biochemical pathway depends on its enzymes. If any enzyme in a pathway fails to work, the end-product will not be produced, just as a failure in an assembly-line process stops output of the product. For instance, albino mutations result from lack of an enzyme necessary for production of the pigment melanin (Figure 4). Stopping a step in a pathway is a useful means to control the output of the cell machinery, so cells contain inhibitors to carry out such control functions, as in the control of melanin production. As another example, the protein that forms blood clots is present in tissues, but in soluble form, and a clot will develop only when a piece is cut off this precursor molecule. The enzyme that cuts this protein is also present, but is normally inactive; when blood vessels are damaged, factors are released that alter the clotting enzyme, so that it immediately becomes active, leading to clotting of the protein.

5. A. The three-dimensional structure of the protein myoglobin (a muscle protein similar to the red blood cell protein haemoglobin), showing the individual amino acids in the protein chain, numbered from 1 to 150, and the iron-containing haem molecule that the protein holds. The haem binds oxygen or carbon dioxide, and the protein's function is to carry these gas molecules.

B. The structure of DNA, the molecule that carries the genetic material in most organisms. It consists of two complementary strands, wound around each other in a helix. The backbone of each strand is formed of molecules of the sugar deoxyribose (S), linked to each other through phosphate molecules (P). Each sugar is connected to a type of molecule called a nucleotide; these form the 'letters' of the genetic alphabet. There are four types of nucleotide: adenine (A), guanine (G), cytosine (C), and thymine (T). A given nucleotide from one strand is paired with a complementary nucleotide from the other, as indicated by the double lines. The rule for this pairing is that A binds to T and G binds to C. When DNA replicates during cell division, the two strands unwind, and a complementary daughter strand is synthesized from each parental strand according to this pairing rule. In this way, a place where A and T bind to each other in the parental molecule produces a place with A and T in each of the daughter molecules.

A

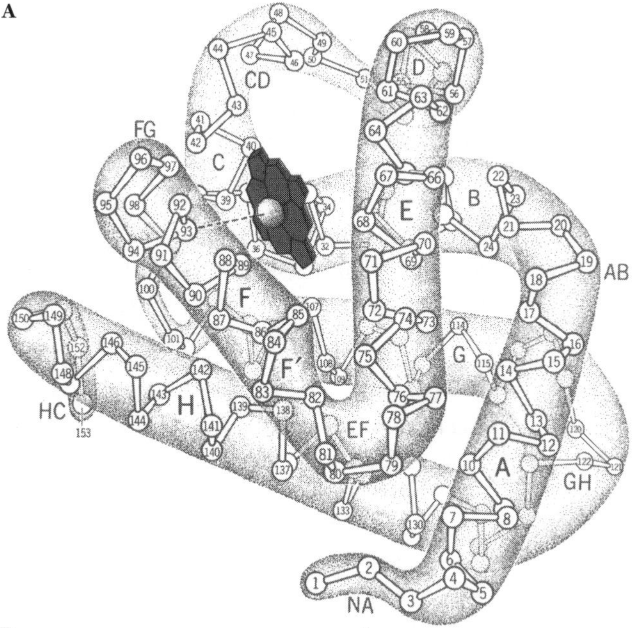

B

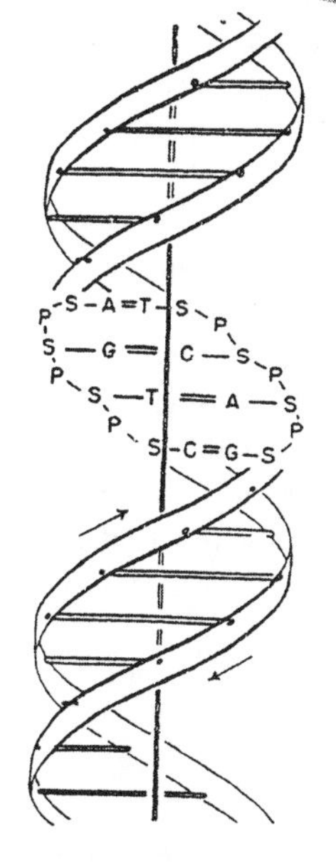

Proteins are very large molecules made up of strings of dozens to a few hundreds of *amino acid* subunits, each joined to a neighbouring amino acid, forming a chain (Figure 5A). Each amino acid is a quite complex molecule, with individual chemical properties and sizes. Twenty different amino acids are used in the proteins of living organisms; a particular protein, such as the haemoglobin in our red blood cells, has a characteristic set of amino acids in a particular order. Given the correct sequence of amino acids, the protein chain folds up into the shape of the working protein. The complex three-dimensional structure of a protein is completely determined by the sequence of amino acids in its constituent chain or chains; in turn, this sequence is completely determined by the sequence of chemical units of the *DNA* (Figure 5B) of the gene that produces the protein, as we will soon explain.

Studies of the three-dimensional structures of the same enzyme or protein in widely different species show that these are often extremely similar across huge evolutionary distances, such as between bacteria and mammals, even if the sequence of amino acids has changed greatly. An example is the myosin protein that we have already mentioned, which is involved in signalling in flies' eyes and in mammalian ears. Such fundamental similarities mean that, astonishingly, it is often possible to correct a metabolic defect in yeast cells by introducing a plant or animal gene with the same function. Yeast cells with a mutation causing a defect in ammonium uptake have been 'cured' by expressing a human gene in their cells (the gene for the Rhesus blood-group protein, RhGA, which was suspected to have the relevant function). The natural (non-mutant) yeast version of this protein has many amino acid differences from the human RhGA one, yet in this experiment the human protein can function in yeast cells lacking their own normal version. The result of this experiment also tells us that a protein with an altered amino acid sequence can sometimes work quite well.

The basis of heredity is common to all organisms

The physical basis of inheritance is fundamentally similar in all eukaryote organisms (animals, plants, and fungi). Our understanding of the mechanism of inheritance, that is the control of individuals' many different characteristics by physical entities that we now call *genes*, first came from work by Gregor Mendel on

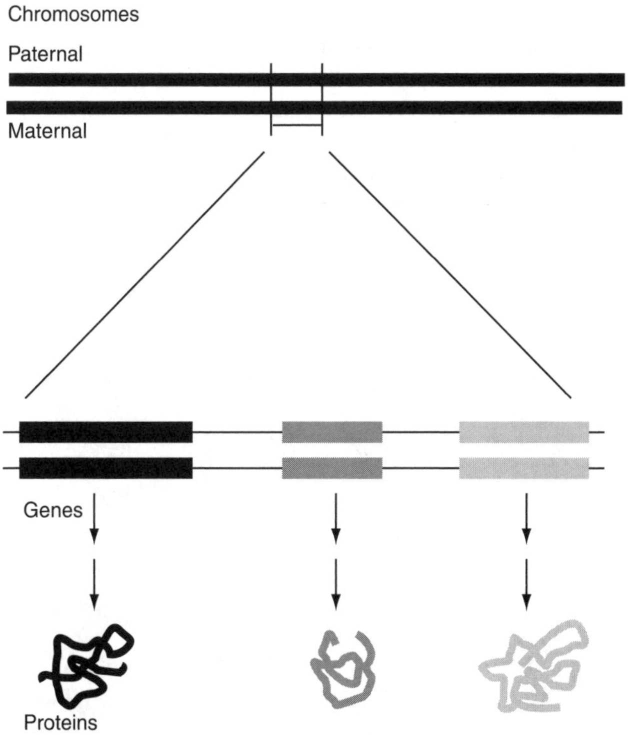

Chromosomes

Paternal

Maternal

Genes

Proteins

6. Diagram of one pair of chromosomes, with a schematic drawing of a small region magnified to show three genes that are located in this chromosome region, and the non-coding DNA in between them. The three different genes are shown as different shades of grey, to indicate that each gene encodes a different protein. In a real cell, only some of the proteins would be produced, while other genes would be turned off so that their proteins would not be formed.

garden peas, but the same rules of inheritance apply to other plants and to animals, including humans. The genes that control the production of metabolic enzymes and other proteins (and thus determine individuals' characteristics) are stretches of DNA carried in the *chromosomes* of each cell (Figures 6 and 7). The discovery that the chromosomes carry the organism's genes in a linear arrangement was first made in the fruitfly, *Drosophila melanogaster*, but it is equally true for our own genome. The order of genes on the chromosomes can be rearranged during evolution, but changes are infrequent, so that sets of the same genes in the same order can be found in the human genome and in the chromosomes of other mammals such as cats and dogs. A chromosome is essentially a single very long DNA molecule encoding hundreds or thousands of genes. The DNA of a

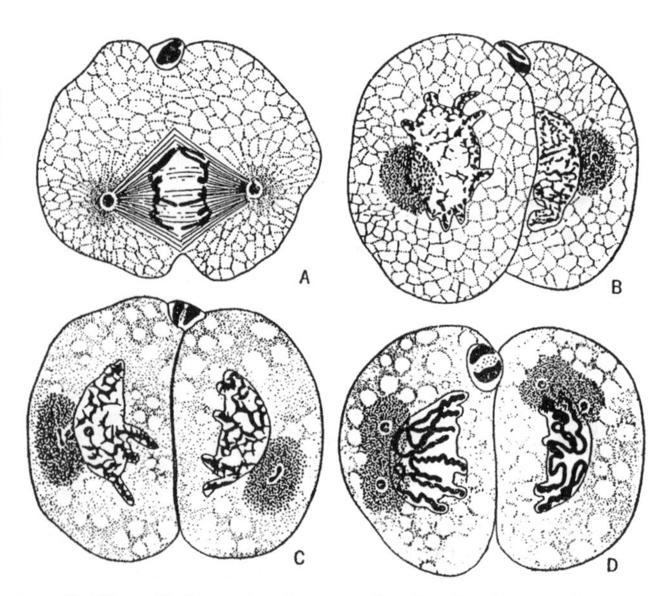

7. A dividing cell of a nematode worm, showing the chromosomes no longer enclosed in the nuclear membrane (A), several stages in the division process (B, C), and finally the two daughter cells, each with a nucleus enclosed in a membrane (D).

chromosome is combined with protein molecules that help to package it in neat coils inside the cell nucleus (resembling the devices used for keeping computer cables tidy).

In higher eukaryotes like ourselves, each cell contains one set of chromosomes derived from the mother through the egg nucleus, and another set derived from the father through the sperm nucleus (Figure 6). In humans, there are 23 different chromosomes in a single maternal or paternal set; in *Drosophila melanogaster*, which is used for much research in genetics, the chromosome number is five (one of which is tiny). The chromosomes carry the information needed to specify the amino acid sequences of an organism's proteins, together with the controlling DNA sequences that determine which proteins will be produced by the organism's cells.

What is a gene, and how does it determine the structure of a protein? A gene is a sequence of the four chemical 'letters' (Figure 5) of the *genetic code*, in which sets of three adjacent letters (*triplets*) correspond to each amino acid in the protein for which the gene is responsible (Figure 8). The gene sequence is 'translated' into the sequence of a protein chain; there are also triplets marking the end of the amino acid chain. A change in the sequence of a gene causes a mutation. Most such changes will lead to a different amino acid being placed in a protein when it is being made (but, because there are 64 possible triplets of DNA letters, and only 20 amino acids used in proteins, some mutations do not change the protein sequence). Across the entire range of living organisms, the genetic code differs only very slightly, strongly suggesting that all life on Earth may have a common ancestor. The genetic code was first studied in bacteria and viruses, but was soon checked and found to be the same in humans. Almost every possible mutation that this code can generate in the sequence of the human red blood cell protein haemoglobin has been observed, but mutations that are impossible with this particular code do not occur.

Human	aac	cag	aca	gga	gcc	cgg	tgc	ctg	gag	gtg	tcc	atc	tct	gac	ggg	ctc	ttc	ctc	agc	ctg
Protein	Asn	Glu	Thr	Gly	Ala	Arg	Cys	Leu	Glu	Val	Ser	Ile	Ser	Asp	Gly	Leu	Phe	Leu	Ser	Leu
Human	aac	cag	aca	gga	gcc	cgg	tgc	ctg	gag	gtg	tcc	atc	tct	gac	ggg	ctc	ttc	ctc	agc	ctg
Chimpanzee	aac	cag	aca	gga	gcc	cgg	tgc	ctg	gag	gtg	tcc	atc	CcT	gac	ggg	ctc	ttc	ctc	agc	ctg
													Pro							
Dog	aac	cag	acC	ggG	Ccc	cgg	tgc	ctg	gag	gtg	tcc	att	CcC	Aac	ggg	ctG	ttc	ctc	agc	ctg
			*	*	Pro								Pro	*		*				
Mouse	aac	cag	Tca	gAG	CcT	Tgg	tgc	ctg	TaT	gtg	tcc	atc	CcA	gaT	ggC	ctc	ttc	ctc	agc	ctA
			Ser	Glu	Pro	Trp			Tyr				Pro	*	*					*
Pig	aac	cag	acG	ggC	Ccc	cAg	tgc	ctg	gag	gtg	tcc	atT	CCC	gac	ggg	ctc	ttc	ctc	agc	ctg
					Pro	Gln						*	Pro							

Human	ggg	ctg	gtg	agc	ttg	gtg	gag	aac	gcg	ctg	gtg	gtg	gcc	acc	atc	gcc	aag	aac	cgg	aac
Protein	Leu	Gly	Leu	Val	Ser	Leu	Val	Glu	Asn	Ala	Leu	Val	Val	Ala	Thr	Ile	Ala	Lys	Asn	Arg
Human	ggg	ctg	gtg	agc	ttg	gtg	gag	aac	gCg	ctg	gtg	gtg	gcc	acc	atc	gcc	aag	aac	cgg	aac
Chimpanzee	ggg	ctg	gtg	agc	ttg	gtg	gag	aac	ATg	ctg	gtg	gtg	gcc	acc	atc	gcc	aag	aac	cgg	aac
									Met											
Dog	ggg	ctg	gtg	agc	Gtt	gtg	gaa	aaT	gTg	ctg	gtg	gtg	gcc	Gcc	atT	gcc	aag	aac	cgC	aa
					*			*	Val					*	*				*	
Mouse	ggg	ctg	gtg	agt	Ctg	gtg	gag	aaT	gTg	ctg	gtT	gtg	ATA	Gcc	atc	Acc	aaa	aac	cgC	aac
				*	*			*	Val		*		*	*					*	
Pig	ggg	ctg	gtg	agC	ctC	gtg	gag	aac	gTg	ctg	gtg	gtg	gcc	Gcc	atc	gcc	aag	aac	cgC	aac
									Val					*					*	

8. DNA and protein sequences of a part of the gene for the melanocyte-stimulating hormone receptor shown in Figure 4, in humans and several other mammals. The figure shows only 40 amino acids out of the total of 951 in the protein. The human DNA sequences are shown at the top, with spaces between the sets of three DNA letters, and the protein sequence is in the grey bars below this (using a three-letter code for the different amino acids). The other species are shown below. Where the DNA sequences differ from the human one, the letter is printed as a capital letter. Triplets that include a difference from the human sequence, but code for the same amino acid as in humans, are asterisked, while triplets that encode differences from the human protein sequence are highlighted. Many red-haired people have an amino acid variant in triplet 151.

In order to produce its protein product, the DNA sequence of a gene is first copied into a 'message' made of a related molecule, *RNA*, whose sequence of 'letters' is copied from that of the gene by a copying enzyme. The RNA message interacts with an elaborate piece of cellular machinery, made up of a conglomeration of proteins and other RNA molecules, to translate the message and produce the protein specified by the gene. This process is essentially the same in all cells, although in eukaryotes it occurs in the cytoplasm, and the message must first move from the nucleus to the cell regions where the translation machinery is sited. In between the genes on the chromosomes are stretches of DNA which do not code for proteins; some of this *non-coding* DNA has the important function of acting as sites for binding proteins that turn the production of the RNA messages of genes on or off as needed. For instance, the genes for haemoglobin are turned on in cells developing into red blood cells, but off in brain cells.

Despite the enormous differences in the modes of life of different organisms, ranging from unicellular organisms to bodies composed of billions of cells with highly differentiated tissues, eukaryote cells undergo similar cell division processes. Single-called organisms such as an amoeba or yeast can reproduce simply by division into two daughter cells. A fertilized egg of a multicellular organism, produced by the fusion of an egg and a sperm, similarly divides into two daughter cells (Figure 7). Many further rounds of cell divisions then take place to produce the many cells and tissue types that form the body of the adult organism. In a mammal, there are over 300 different types of cell in the adult body. Each type has a characteristic structure and produces a specific array of proteins. The arrangement of these cells into tissues and organs during development requires elaborately controlled networks of interactions between the cells of the developing embryo. Genes are turned on and off to ensure that the right kind of cell is produced in the right place at the right time. In some well-studied organisms, such as *Drosophila melanogaster*, we now know a great deal about how these interactions result in the emergence of the intricate body plan of the fly from the apparently undifferentiated egg cell. Many signalling processes involved in development and differentiation of particular tissues, such as nerves, are found to be shared by all multicellular animals, while land plants use a rather different set, as might be expected from the fact that the fossil record shows that multicellular animals and plants have separate evolutionary origins (see Chapter 4).

When a cell divides, the DNA of the chromosomes is first replicated, so that there are two copies of each chromosome. Cell division is a process with tight controls to ensure that the newly copied DNA sequence undergoes 'proof-reading' for errors. Cells have enzymes that, using certain properties of the way DNA is replicated, can distinguish new DNA from the old 'template' DNA. This enables most errors in copying to be detected and corrected, ensuring that the template has been faithfully copied before the cell is allowed to proceed to the next step, division of the cell itself. The machinery of cell division ensures that each daughter cell receives a complete copy of the set of chromosomes that was present in the parent cell (Figure 7).

Most prokaryotes' genes (including those of many viruses) are also sequences of DNA which are organized only slightly differently from those carried in eukaryote chromosomes. Many bacteria have just one circular DNA molecule as their genetic material. Some viruses, however, such as those responsible for influenza and AIDS, have genes made of RNA. The proof-reading that occurs in DNA replication does not happen when RNA is copied, and so these viruses have extremely high mutation rates, and can evolve very rapidly within the host's body. As we will describe in Chapter 5, this means that it is difficult to develop vaccines against them.

Eukaryotes and prokaryotes differ greatly in their amounts of non-coding DNA. The bacterium *Escherichia coli* (a normally harmless species that lives in our intestines) has about 4,300 genes, and the stretches that code for protein sequences form about 86% of this species' DNA. In contrast, less than 2% of the DNA in the human genome codes for protein sequences. Other organisms lie between these extremes. The fruitfly, *Drosophila melanogaster*, has about 14,000 genes in about 120 million 'letters' of DNA, and about 20% of the DNA is made up of coding sequences. The number of different genes in the human genome is still not precisely known. The current best count comes from the sequencing of the complete genome. This allows geneticists to recognize sequences that are probably genes, based on what we know from genes that had previously been studied. It is a difficult task to find these sequences in the huge amount of DNA that makes up the genome of any species, particularly for our own genome, which has a very large DNA content (25 times as much as the fruitfly). The number of genes in humans is about 35,000, much smaller than had been guessed from the number of cell and tissue types with different functions. The number of proteins a human can make is probably considerably larger than this, because this method of counting cannot detect very small genes, or unconventional ones (for example, genes that lie within other genes, which exist in several organisms). It is not yet known how much of the non-coding DNA is important for the life of the organism. Although much of it is made up of viruses and other parasitic entities that live in chromosomes, some of it has important functions. As we have already mentioned, there are DNA sequences outside genes that can bind proteins controlling which genes in a cell are 'turned on'. The control of gene activity must be much more important in multicellular creatures than in bacteria.

In addition to the discovery that widely different organisms have DNA as their genetic material, modern biology has also uncovered profound similarities in the life-cycles of eukaryotes, despite their diversity, which ranges from unicellular fungi such as yeasts, to annual plants and animals, to long-lived (though not immortal) creatures like ourselves and many trees. Many, though not all, eukaryotes have a sexual stage in each generation, in which the maternal and paternal genomes of the uniting egg and sperm (each made up of a set of some number n of different chromosomes, characteristic of the species in question) combine to make an individual with $2n$ chromosomes. When an animal makes new eggs or sperm, the n condition is restored by a special kind of cell division. Here, each pair of maternal and paternal chromosomes lines up, and (after exchanging material to form chromosomes that are patchworks partly of paternal and partly of maternal DNA) the chromosome pairs separate from each other in a similar way to the separation of newly replicated chromosomes in other cell divisions. At the end of the process, the number of chromosomes in each egg or sperm cell nucleus is therefore halved, but each egg or sperm has one complete set of the organism's genes. The double set will be restored on the union of egg and sperm nuclei at fertilization.

The basic features of sexual reproduction must have evolved long before the evolution of multicellular animals and plants, which are latecomers on the evolutionary scene. This is clear from the common features displayed in the reproduction of sexual unicellular and multicellular organisms, and the similar genes and proteins that have been discovered to be involved in the control of cell division and chromosome behaviour in groups as distant as yeast and mammals. In most single-celled eukaryotes, the $2n$ cell produced by fusion of a pair of cells, each with n chromosomes, divides immediately to produce cells with n chromosomes, just as described above for germ cell production in multicellular animals. In plants, the reduction of chromosome number from $2n$ to n happens before egg and sperm formation, but the same kind of special cell division is again involved; in mosses, for instance, there is a prolonged life-cycle stage with chromosome number n that forms the moss plant, on which the small $2n$ parasitic stage develops after eggs and sperm are made and fertilization has occurred.

The complications of such sexual processes are absent from some multicellular organisms. In such 'asexual' species, mothers produce daughters without a reduction of chromosome number from $2n$ during egg production. Nevertheless, all multicellular asexual organisms show clear signs of being descended from sexual ancestors. For example, common dandelions are asexual; their seeds form without the need for pollen to be brought to the flowers, as is required for most plants to reproduce. This is an advantage to a weedy species like the common dandelion, which speedily generates large numbers of seeds, as anyone who has a lawn can see for themselves. Other dandelion species reproduce by normal matings between individuals, and common dandelions are so closely related to these that they still make pollen that can fertilize the flowers of the sexual species.

Mutations and their effects

Despite the proof-reading mechanisms that correct errors when DNA is copied during cell division, mistakes do occur, and these are the source of mutations. If a mutation results in a change in the amino acid sequence of a protein, the protein may malfunction; for example, it may not fold up correctly and so may be unable to do its job properly. If it is an enzyme, this can cause the metabolic pathway to which it belongs to run slowly, or not at all, as in the case of the albino mutations already mentioned. Mutations in structural or communication proteins may impair cell functions or the organism's development. Many diseases in humans are caused by such mutations. For instance, mutations in genes involved in controlling cell division increase the risk of cancer developing. As already mentioned, cells have exquisite control systems to ensure that they divide only when everything is in order (proof-reading for mutations must be complete, the cell must show no signs of infection or other damage, and so on). Mutations affecting these control systems can result in uncontrolled cell division, and malignant growth of the cell lineage. Luckily, it is unusual for both members of a pair of genes in a cell to be mutant, and one non-mutant member of the pair is often enough for correct cell functioning. A cell lineage also usually requires other adaptations to become a successful cancer, so malignancy is uncommon. (A blood supply is needed for tumours, and the cells' abnormal characteristics must evade detection by the body.) Nevertheless, understanding cell division and its control is a major part of cancer research. The process is so similar in cells of different eukaryote organisms that the 2001 Nobel prizes in medicine were given for research on cell division in yeast, which showed that a gene involved in the control system of yeast cells is mutated in some human familial cancers.

Mutations that give a predisposition to cancer are rare, as are most other disease-causing mutations. The most common genetic disorder in northern European human populations is cystic fibrosis, but even in this case the non-mutant sequence of the gene involved represents more than 98% of copies of the gene in the population. Mutations that cause failure of an important enzyme or protein may lower the survival or fertility of affected individuals. The gene sequence that leads to the non-functional enzyme will thus be under-represented in the next generation, and will eventually be eliminated from the population. A major role of natural selection is to keep the proteins and other enzymes of most individuals working well. We will revisit this idea in Chapter 5.

One important type of mutation leads to a protein not being produced in sufficient amounts by its gene. This could happen because of a problem in the normal control system for that gene, which either does not switch it on when it should do so, does not produce in the right quantities, or stops production of the protein before it is finished. Other mutations may not abolish an enzyme's production, but the enzyme may be faulty, just as a production line can be hindered or stopped if one of the necessary tools or machines is defective in some way. If one or more of the component amino acids are missing, the protein may not function correctly, and the same can happen if a different amino acid appears at a particular position in the chain, even if all the rest are correct. Mutations causing loss of function can contribute to evolution when selection no longer acts to eliminate them (see Chapters 2 and 6 for how selectively neutral mutations can spread). About 65% of human olfactory receptor genes are 'vestigial genes' that do not produce working receptor proteins, so we have many fewer olfactory functions than mice or dogs (not surprisingly, given the importance of smell in their daily lives and social interactions, compared with its minor role in ours).

There are also many differences between normal individuals in a species. For instance, individuals in human populations differ in their ability to taste or smell certain chemicals, or to break down some chemicals used as anaesthetics. People who lack an enzyme that breaks down an anaesthetic may suffer a bad reaction to it, but the lack of the enzyme would otherwise not matter. Similar differences in the ability to deal with other drugs, and sometimes foods, are an important aspect of variability in humans, and knowledge of these differences is necessary for modern medicine, in which strong drugs are often used.

Mutations in the enzyme glucose-6-phosphate dehydrogenase (an enzyme for an early step in the pathway by which cells derive energy from glucose) illustrate some of these kinds of differences. Individuals entirely missing this gene cannot survive (because the pathway in which it functions is vital in controlling the levels of toxic chemicals produced as a by-product of cellular energy generation). In human populations, there are at least 34 different normal variants of the protein that are not only compatible with healthy life, but are actually protective against malaria parasites. Each of these differs by one or a few amino acids from the protein's most common normal sequence. Several of these variants are widespread in Africa and the Mediterranean regions, and in some malarial populations variant individuals are frequent. However, some of the variants cause a form of anaemia when a type of bean is eaten, or when certain anti-malarial drugs are given. The well known ABO and other blood-groups are another example of normal variability within the human population; they are due to variation in the sequences of proteins that control details of the surfaces of red blood cells. Variation in the receptor protein for melanocyte-stimulating hormone, which is important in the production of the skin pigment melanin (see Figure 4), can cause hair colour differences; in many red-haired people, this protein has an altered amino acid sequence. As we shall discuss in Chapter 5, genetic variability is the essential raw material on which natural selection acts to produce evolutionary changes.

Biological classification and DNA and protein sequences

A new and important set of data providing clear evidence that organisms are related to one another through evolution comes from the letters in their DNA, which can now be "read" by the chemical procedure of DNA sequencing. Systems of biological classification based on visible characteristics, which were developed over the past three centuries of study of plants and animals, are now supported by recent work comparing DNA and protein sequences among different species. Measuring the similarity of DNA sequences makes it possible to have an objective concept of relationship among species. We will describe this in more detail in Chapter 6. For the moment we need only understand that the DNA sequences of a given gene will be most similar for more closely related species, while those of more distantly related species are more different (Figure 8). The amount of difference increases roughly proportionally to the amount of time separating two sequences being compared. This property of molecular evolution allows evolutionary biologists to estimate times of events that cannot be studied in fossils, using a *molecular clock*. For instance, we have already mentioned changes in the order of an organism's genes on its chromosomes. A molecular clock can be used to estimate the rate of such chromosomal rearrangements. Consistent with the evolutionary viewpoint, species that we believe to be close relatives, such as humans and rhesus monkeys, have chromosomes that differ by fewer rearrangements than humans and New World primates such as the woolly monkey.

In the next chapter, we will explain the evidence for evolution based on fossil data, and from data on the geographical distribution of living species. These observations complement those described here, in showing that the theory of evolution provides a natural explanation for a wide range of biological phenomena.

Chapter 4
The evidence for evolution: patterns in time and space

> The history of man, therefore, is but a short ripple in the ocean of time.
>
> From *On the Interaction of the Natural Forces*,
> Hermann von Helmholtz, 1854

The age of the Earth

It would have been impossible to realize that living organisms have originated by evolution, without the success of late 18th- and early 19th-century geologists in establishing that the present-day structure of the Earth is itself the product of long-continued physical processes. The methods involved are similar in principle to those used by historians and archaeologists. As the great French naturalist, the Comte de Buffon, wrote in 1774:

> Just as in civil history we consult warrants, study medallions, and decipher ancient inscriptions, in order to determine the epochs of human revolutions and fix the date of moral events, so in natural history one must dig through the archives of the world, extract ancient relics from the bowels of the earth, gather together their fragments and assemble again in a single body of proofs all those indications of the physical changes which can carry us back to the different Ages of Nature. This is the only way of fixing certain points in the immensity of space, and of placing milestones on the eternal path of time.

At the risk of some oversimplification, there were two key insights that led to the successes of early geology; the principle of *uniformitarianism*, and the invention of *stratigraphy* as a method of dating. Uniformitarianism is particularly associated with the late 18th-century Edinburgh geologist James Hutton, and was codified later by another Scottish scientist, Charles Lyell, in his *Principles of Geology* (1830). It is simply the application to the history of the structure of the Earth of the same principle used by astronomers in attempting to understand the constitution of distant planets and stars: the basic physical processes involved are assumed to be the same everywhere and at all times. Geological change over time reflects the operation of the laws of physics, which are themselves unchanging. For example, physical theory implies that the speed of rotation of the Earth must have decreased over millions of years because of frictional forces induced by the tides, which are caused by the gravitational forces of the Sun and Moon. The length of the day is now much longer than when the Earth was first formed, but the strength of the force of gravity has not changed.

There is, of course, no independent justification of this assumption of uniformity, any more than there is any logical justification for the assumption of the regularity of nature that underlies the most basic aspects of our daily life. Indeed, there is no distinction between these two assumptions, except the scale of time and space to which they apply. Their justification is that, first, uniformitarianism represents the simplest possible basis on which we can proceed to interpret events that are remote in time and space. Second, it has proved to be remarkably successful.

The uniformitarian assumption in geology implies that the present-day constitution of the Earth's surface reflects the cumulative action of processes of formation of new rocks by volcanic action and deposition of sediments in rivers, lakes, and seas, and the erosion of old rocks by the action of wind, water, and ice. The formation of *sedimentary* rocks like sandstone and limestone depends on the erosion of other rocks. In contrast, the formation of mountains by volcanic action and uplift of land by earthquakes must precede their degradation by erosion. These processes can be observed in action in the present day; anyone who has visited a mountainous region, especially at a time of year when freezing and thawing is happening, will have witnessed erosion of rocks, and the transport of the resulting debris downstream by streams and rivers. The deposition of sediments at the mouths of rivers is also easy to observe. Volcanic and earthquake activity are confined to certain regions of the globe, especially the edges of continents and middles of oceans, for reasons which are now well understood, but there are numerous recorded instances of the formation of new oceanic islands by volcanic action, and of the uplift of land by earthquakes. In *The Voyage of the Beagle*, Darwin described the effects of the Chilean earthquake of February 1835 in the following terms:

> The most remarkable effect of this earthquake was the permanent elevation of the land; it would probably be more correct to speak of it as the cause. There can be no doubt that the land around the Bay of Concepcion was upraised two or three feet ... At the island of S. Maria (about thirty miles distant) the elevation was greater; on one part Captain Fitzroy found beds of putrid mussel-shells *still adhering to the rocks*, ten feet above high water mark ... The elevation of this province is particularly interesting from its having been the theatre of several other violent earthquakes, and from the vast number of sea-shells scattered over the land, up to a height of certainly 600, and I believe, 1000 feet.

Geology has been outstandingly successful in interpreting the structure of the Earth at or near its surface in terms of these processes, and in reconstructing the events that have led to the present-day appearance of many parts of the Earth. The order of these events can be established by the principle of stratigraphy. Information on the mineral composition and arrays of fossils found in different layers of rocks (*strata*) is used to characterize individual layers. The recognition that fossils represent the preserved remains of long-dead plants and animals, rather than artefacts of mineral formation, was essential for the success of stratigraphy. The types of fossils found in a given sedimentary rock layer provide evidence about the environment that prevailed when it was laid down; for example, it is usually possible to tell whether the organisms were marine, freshwater, or terrestrial. Fossils are, of course, absent from rocks such as granite or basalt that form by the solidification of molten material from below the Earth's crust.

During his travels throughout Britain to construct canals in the early 19th century, the English canal engineer William Smith recognized that similar successions of strata occur in different parts of Britain (which has an unusual variety of rocks of different ages for such a small area of land). Using the principle that older rocks must normally lie below younger ones, the comparison of the succession of strata in different localities enabled geologists to reconstruct sequences of strata that were laid down through immense periods of time. If rocks of type A are found below type B in one location, and B is found below C somewhere else, then one infers the sequence A-B-C, even if A and C are never found together in one place.

The systematic use of this method by 19th-century geologists allowed them to determine the major divisions of geological time (Figure 9). These yield a relative, not an absolute, chronology; absolute dates require methods for calibrating the rate of the processes involved, which is very difficult to do with any precision. The processes involved in landscape formation are very slow; erosion of even a few millimetres of rock takes many years, and the deposition of sediments is correspondingly slow. Similarly, uplift of land even in the most active mountain-building areas, such as the Andes, occurs at a rate of only a fraction of a metre a year on average. Given the existence of sedimentary rocks of the same formation that are several kilometres deep in many parts of the

Era	Period	Epoch	years ago
	Quaternary	Holocene	10,000
		Pleistocene	2 million
Cenozoic	Tertiary	Pliocene	7 million
		Miocene	26 million
		Oligocene	38 million
		Eocene	54 million
		Palaeocene	64 million
Mesozoic	Cretaceous		136 million
	Jurassic		190 million
	Triassic		225 million
Palaeozoic	Permian		280 million
	Carboniferous		345 million
	Devonian		410 million
	Silurian		440 million
	Ordovician		530 million
	Cambrian		570 million

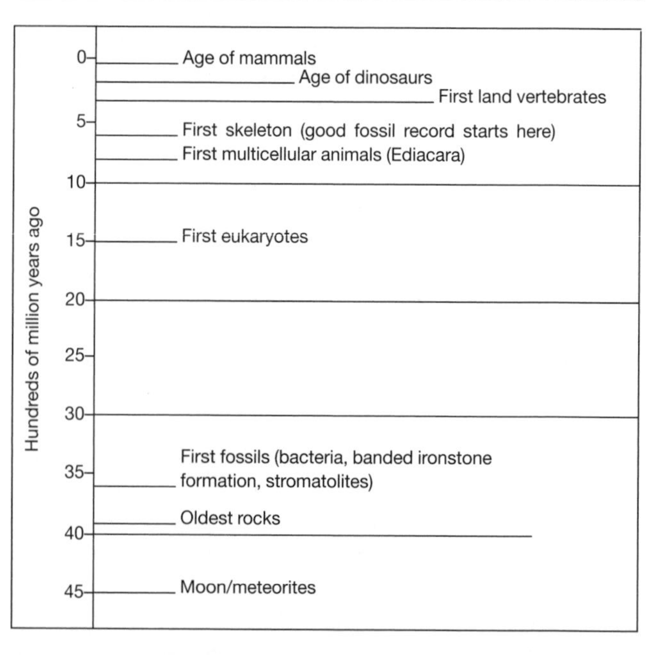

world, and the evidence that equally large deposits have been eroded, the necessity of a time-scale of at least many tens of millions of years for the existence of the Earth was quickly recognized, in conflict with Biblical chronology. Lyell suggested on this basis that the Tertiary period lasted about 80 million years, and that the Cambrian occurred 240 million years ago.

Such a long time-scale for the Earth was challenged by the eminent physicist Lord Kelvin, on the grounds that the rate of cooling of an originally molten Earth would make the Earth's core much cooler than it actually is, if the Earth had been formed more than approximately 100 million years ago. Kelvin's calculation was correct for the physics of his day. However, at the end of the 19th century, radioactive decay of unstable elements, such as uranium, into more stable derivatives was discovered. This process of decay is accompanied by the release of energy sufficient to slow the rate of cooling of the Earth to a value that agrees with current estimates of its age.

Radioactivity also provided new and reliable methods for establishing the ages of rock samples. The atoms of radioactive elements have a constant probability per year of decaying to a more stable daughter element, accompanied by the emission of radiation. When a rock is laid down, it can be assumed that the element in question is pure; hence, if the proportion of the daughter element in the sample is measured, the time since the formation of the rock can be estimated, knowing the rate of the decay process as established by experiments. Different elements are useful for dating rocks of different ages. Determinations of the ages of rocks belonging to the different periods of geology by this technique have given us the

9. The major divisions of geological time. The upper part shows the named divisions from the Cambrian onwards, in which most fossils are found (this is less than one-eighth of the time since the formation of the Earth). The lower part shows the major events that have occurred during Earth history.

dates accepted today. While the methods are constantly being refined, and the dates revised, the general time-scale that they indicate is very clear (Figure 9). It establishes an immense, almost incomprehensible, amount of time for biological evolution to occur.

The fossil record

The fossil record is our only direct source of information on the history of life. To interpret it correctly, it is necessary to understand how fossils are formed, and how scientists study them. When a plant, animal, or microbe dies, the soft parts are almost certain to decay rapidly. Only in unusual environments, such as the arid atmosphere of a desert or the preservative chemicals of a piece of amber, are the microbes responsible for decay unable to break down the soft parts. Remarkable cases of preservation of soft parts, sometimes going back tens of millions of years in the case of insects trapped in amber, have been found, but these are the exception rather than the rule. Even skeletal structures, such as the tough chitin which covers the bodies of insects and spiders, or the bones and teeth of vertebrates, eventually decay. Their slower rate of disappearance offers, however, an opportunity for minerals to infiltrate them, and eventually replace the original material with a mineralized replica (occasionally this happens to soft parts as well). Alternatively, they may create a mould of their shape as minerals are deposited around them.

Fossilization is most likely to happen in aquatic environments, where the deposition of sediment and precipitation of minerals occur at the bottom of seas, lakes, and river estuaries. Remains that sink to the bottom can then turn into fossils, although the chance that this happens for a given individual is extremely small. The fossil record is therefore very biased: marine organisms living in shallow seas, where sediments are continuously formed, have the best fossil record, and flying creatures have the worst. In addition, the deposition of sediments may be interrupted, for example by a change in climate or by uplift of the seabed. For many types of creature, we have almost no fossil record; for others, the record is interrupted many times.

An excellent example of the problems caused by this incompleteness is provided by the coelacanth. This is a type of bony fish with lobed fins, related to the ancestors of the first land vertebrates. Coelacanths were abundant in the Devonian era (400 million years ago), but subsequently declined in number. The last fossil coelacanths are dated to about 65 million years ago, and the group was long thought to be extinct. In 1939 fishermen from the Comoro islands off the south-eastern coast of Africa caught a strange-looking fish, which turned out to be a coelacanth. Scientists have subsequently been able to study the habits of living coelacanths, and a new population has been discovered in Indonesia. The coelacanths must have existed continuously over a vast stretch of time, but left no fossil record because of their low abundance and the great depth at which they live.

The gaps in the fossil record mean that it is rare to have a long-continued series of remains showing the more or less continuous changes which are expected under the hypothesis of evolution. In most cases, new groups of animals or plants make their first appearance in the fossil record without any obvious links to earlier forms. The most famous example is the 'Cambrian explosion', which refers to the fact that most of the major groups of animals appear for the first time as fossils in the Cambrian period, between 550 and 500 million years ago (this will be discussed again in Chapter 7).

Nevertheless, as Darwin argued eloquently in *The Origin of Species*, the general features of the fossil record provide strong evidence for evolution. The discoveries of palaeontologists since his day have reinforced his arguments. In the first place, many examples of intermediate forms have been discovered, connecting groups that were formerly thought to be separated by unbridgeable gaps. The fossil bird-reptile *Archaeopteryx*, discovered shortly after the publication of *The Origin of Species*, is perhaps the most famous of these. *Archaeopteryx* fossils are rare (only six specimens exist). They come from Jurassic limestone from about 120 million years ago that was laid down in a large lake in Germany. These creatures show a mosaic of characteristics, some resembling those of modern birds, such as wings and feathers, and others like those of reptiles, such as a toothed jaw (instead of a beak) and a long tail. Many details of their skeletons are indistinguishable from those of a contemporary group of dinosaurs, but *Archaeopteryx* differs from them, as it could clearly fly. Other fossils linking birds and dinosaurs have subsequently been found, and it has recently been shown that dinosaurs with feathers existed before *Archaeopteryx*. Other important intermediates include fossil mammals from the Eocene (about 60 million years ago), with forelimbs and reduced hindlimbs adapted to swimming. These link modern whales to animals that belong to the group of cloven-hoofed herbivores that includes cows and sheep.

Humans are an excellent example of gaps in the record being filled as more research is done. No fossil remains connecting apes and humans were known at the time of publication of Darwin's 1871 book on human evolution, *The Descent of Man*. Darwin argued on the basis of anatomical similarities that humans were most closely related to gorillas and chimpanzees, and had therefore probably originated in Africa from an ancestor that also gave rise to these apes. A whole series of fossil remains have since been found and accurately dated by the methods described earlier, and new fossils continue to be found. The nearer in time to the present, the more similar are the fossils to modern humans (Figure 10); the earliest fossils that can be assigned clearly to *Homo sapiens* date from only a few hundred thousand years ago. In agreement with Darwin's inferences, early human evolution probably took place in Africa, and it seems likely that our relatives first entered Eurasia about 1.5 million years ago.

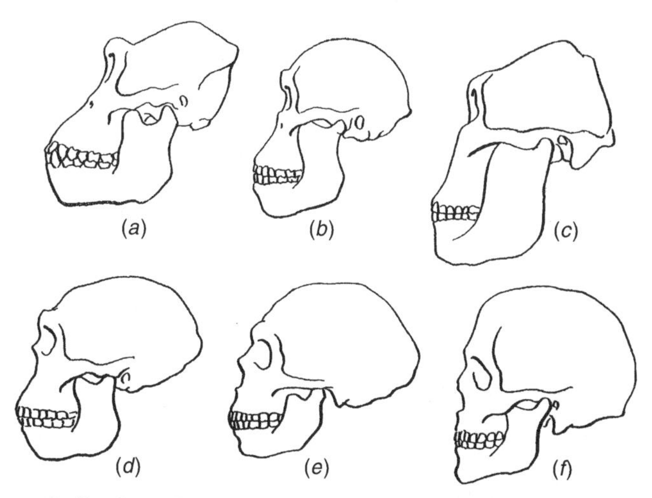

10. Skulls of some human ancestors and relatives. A. Female gorilla. B and C. Fossils of two different species of one of the earliest human relatives, *Australopithecus*, from about 3 million years ago. D. Fossil of an intermediate between *Australopithecus* and modern humans called *Homo erectus*, from about 1.5 million years ago. E. A fossil Neanderthal human, *Homo neanderthalensis*, from about 70,000 years ago. F. Modern human, *Homo sapiens*.

There are also cases of almost completely continuous temporal sequences of fossils, in which it seems certain that we have a record of change in a single evolving lineage. The best examples come from studies of the results of drilling down into deposits at the bottom of the sea, from which long rock columns can be recovered. This allows very fine-scaled time separation between successive samples of the microorganisms whose innumerable fossilized skeletons form the body of the rock. Careful measurements of the shapes of the skeletons of creatures such as foraminiferans, which are single-celled marine animals, allows characterization both of the averages and levels of variability of successive populations over a long period of time (Figure 11).

Even if there were no graded intermediates in the fossil record, the general features of the record are barely comprehensible except in

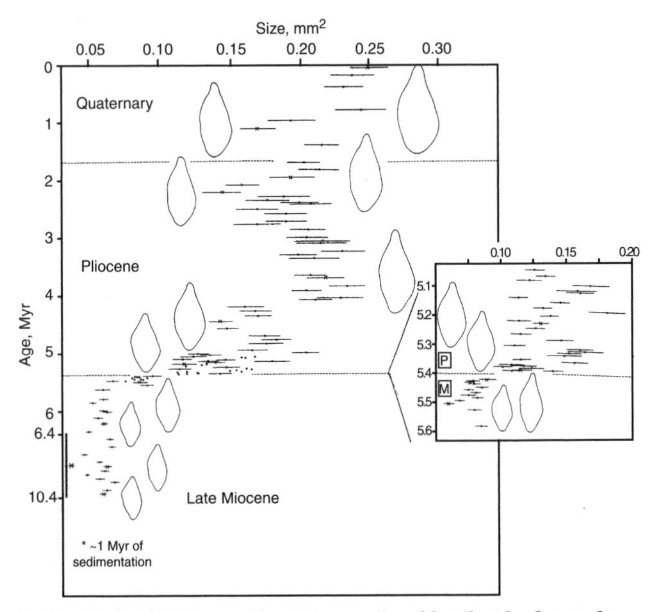

11. Gradual evolutionary change in a series of fossils. The figure shows the means and ranges of body sizes in samples of a fossil foraminiferan (*Globorotalia tumida*), a single-celled marine shelled animal. Size changes gradually in this lineage, except for two apparent discontinuities. At the boundary between the late Miocene and Pliocene eras, a more detailed set of fossils (inset) shows that the discontinuity observed with the coarser set of fossils almost entirely reflects an episode of very rapid change, since the ranges of most successive samples overlap each other. For the discontinuity just before 4 million years ago, there are currently no fossil data.

the light of evolution. Although the fossil record before the Cambrian era is fragmentary, there is evidence for the remains of bacteria and related single-celled organisms going back more than 3.5 billion years. Much later on, there are remains of more advanced (eukaryote) cells, but still no evidence for multicellular organisms. Organisms made up of simple clusters of cells appear only about 800 million years ago (MYA), at a time of environmental crisis when the Earth was largely covered with ice. About 700–550 MYA, there is evidence for soft-bodied, multicellular animal life.

As already mentioned, animal remains associated with hard skeletons only become abundant in the Cambrian rocks, about 550 MYA. By the end of the Cambrian, around 500 MYA, there is evidence for nearly all major animal groups, including primitive fish-like vertebrates that lack jaws, resembling modern lampreys.

All life until this time is associated with marine deposits, and the only plant remains are algae, which lack the vessels that multicellular land plants use for fluid transport. By 440 MYA, there is evidence for freshwater life, followed by fossil spores that imply the existence of the first land plants; shark-like fish with jaws appear in the sea. In the Devonian (400–360 MYA), freshwater and land remains become much more common and diverse. There is evidence for primitive insects, spiders, mites, and millipedes, as well as simple vascular plants and fungi. Jawed fish with bony skeletons become common, including lobe-finned fishes similar in structure to the first salamander-like amphibians that appear at the end of the Devonian. These are the first land vertebrates.

During the next division of the geological record, the Carboniferous (360–280 MYA), land life-forms become abundant and diverse. The coal deposits, which give this period its name, are the fossilized remains of tree-like plants that grew in tropical swamps, but these are similar to contemporary horse-tails and ferns and are unrelated to modern conifers or deciduous trees. Remains of primitive reptiles, the first vertebrates to become fully independent of water, are found at the end of the Carboniferous. In the Permian (280–250 MYA), there is a great diversification of reptiles; some of these have anatomical features that increasingly come to resemble those of mammals (the mammal-like reptiles). Some of the modern groups of insects, such as bugs and beetles, appear.

The Permian ends with the largest set of extinctions seen in the fossil record, in which some previously dominant groups such as trilobites disappear completely, and many other groups are nearly wiped out. In the recovery that follows, a variety of new forms appear, both on the land and in the sea. Plants similar to modern conifers and cycads appear in the Triassic (250–200 MYA). Dinosaurs, turtles, and primitive crocodiles appear; right at the end, the first true mammals are found. These are distinguished from their precursors by having a lower jaw consisting of a single bone connected to the skull directly (the three bones that form this connection in reptile skulls have evolved into the small internal bones of the mammalian ear, see Chapter 3, p. 15). Bony fishes similar to modern forms appear in the sea. In the Jurassic (200–140 MYA), the mammals diversify somewhat, but life on land is still dominated by the reptiles, especially dinosaurs. Flying reptiles and *Archaeopteryx* appear. Flies and termites appear for the first time, as do crabs and lobsters in the sea. Only in the Cretaceous (140–65 MYA) did flowering plants evolve – the last major group of organisms to evolve. All major modern groups of insects are found by this time. Pouched mammals (marsupials) appear in the middle of the Cretaceous, and forms similar to modern placental mammals are found towards its end. Dinosaurs are still abundant, though in decline at the end of the period.

The Cretaceous ends with the most famous of the major extinction events, associated with the impact of an asteroid that landed in the Yucatan peninsula of Mexico. All the dinosaurs (except birds) disappear, along with many other forms once common on the land and in the sea. This is followed by the Tertiary period, which extends until the beginning of the great Ice Ages, about 2 MYA. During the first divisions of the Tertiary, between 65 and 38 MYA, the main groups of placental mammals appear. At first, these are mostly similar to modern insectivores such as shrews, but some become fairly distinct by the end of this period (whales and bats are recognizable, for instance). Most of the main groups of birds are found, as well as modern types of invertebrates, and all major flowering plant groups except grasses. Bony fishes of an essentially modern type are abundant. Between 38 and 26 MYA, grasslands appear, associated with grazing horse-like animals with three toes (instead of the single toe of modern horses). Primitive apes also appear. Between 26 and 7 MYA, prairie grasslands are widespread in North America, and horses with short lateral toes and high-crowned teeth adapted for grazing are found. Various ungulates, such as pigs, deer, and camels appear, together with elephants. Apes and monkeys become more diverse, especially in Africa. Between 7 and 2 MYA, marine life has an essentially modern aspect, although many forms living then are now extinct. The first remains of creatures with some distinctively human features appear in this period. The end of the Tertiary, from 2 MYA to 10,000 years ago, sees a series of Ice Ages. Most animals and plants are essentially modern in form. Between the end of the last Ice Age 10,000 years ago, and the present, humans become the dominant land animal, and many large mammal species become extinct. There is some fossil evidence for evolutionary change over this period, such as the evolution of dwarf forms of various large mammal species on islands.

The fossil record thus suggests that life originated in the sea over 3 billion years ago, and that for more than a billion years only single-celled organisms related to bacteria existed. This is exactly what is expected on an evolutionary model; the evolution of the machinery needed to translate the genetic code into protein sequences, and the complex organization of even the simplest cell, must have required many steps, the details of which almost defy our imagination. The late appearance in the record of clear evidence for eukaryote cells, with their substantially more complex organization compared with prokaryotes, is also consistent with evolution. The same applies to multicellular organisms, whose development from a single cell requires elaborate signalling mechanisms to control growth and differentiation: these could not have evolved before single-celled forms existed. Once simple multicellular forms evolved, it is understandable that they rapidly diversified into numerous forms, adapted to different modes of life, as occurred in the Cambrian. We shall discuss adaptation and diversification in the next chapter.

The fact that life was exclusively marine for an immense period is also understandable from an evolutionary perspective. Early in the Earth's history, the geological evidence shows that there was very little oxygen in the atmosphere. The consequent lack of protection from ultra-violet radiation by atmospheric ozone, which is formed from oxygen, would have prohibited life on land or even in fresh water. Once sufficient oxygen had built up as a result of the photosynthetic activities of early bacteria and algae, this barrier was removed, and the possibility of the invasion of the land opened up. There is evidence for an increase in atmospheric oxygen levels during the period leading up to the Cambrian, which may have permitted the evolution of larger and more complex animals. Similarly, the appearance of fossils of flying insects and vertebrates after the emergence of life on land makes sense, since it is unlikely that true flying animals could evolve from purely aquatic forms.

The recurrent phenomenon of the emergence of abundant and diverse forms of life, followed by their wholesale extinction (as with the trilobites and dinosaurs) or their reduction to just one or a few surviving forms (like the coelacanths) also makes sense in terms of evolution, whose mechanisms have no foresight and cannot guarantee that their products can survive sudden large environmental changes. Similarly, the rapid diversification of groups after the colonization of a new habitat (as in the invasion of the land), or after the extinction of a dominant rival group (as with the mammals after the disappearance of the dinosaurs), is expected on evolutionary principles.

The interpretation of the fossil record in terms of biological knowledge therefore follows the same principle of uniformitarianism that is applied by geologists to the history of the structure of the Earth. The fossil evidence might have shown patterns that falsify evolution. The great evolutionist and geneticist J. B. S. Haldane is alleged to have answered the question of what observation would cause him to abandon his belief in evolution by saying: 'A pre-Cambrian rabbit'. So far, no such fossil has been found.

Patterns in space

Another important body of facts that makes sense only in terms of evolution comes from the distribution of living creatures over space rather than time, as described by Darwin in two of the fifteen chapters of the *The Origin of Species*. One of the most striking examples of this involves the flora and fauna of oceanic islands, such as the Galapagos and Hawaiian islands, which geological evidence shows were formed by volcanic action and were never connected to a continent. According to the theory of evolution, the present-day inhabitants of such islands must be the descendants of individuals who were able to cross the vast distances separating the newly formed islands from the nearest inhabited land. This puts several restrictions on what we are likely to see. First, the difficulty of colonization of a remote piece of newly formed land means that few species will be able to establish themselves. Second, only types of organism that have characteristics that enable them to cross hundreds or thousands of miles of ocean can become established. Third, even in the groups that are represented, there will be a highly random element to which species are present, because of the small number of species that arrive on the islands. Finally, evolution on such remote islands will produce many forms that are found nowhere else.

These expectations are strikingly verified by the evidence. Oceanic islands do indeed tend to have relatively few species in any major group, compared with continents or offshore islands with comparable climates. The types of organisms found on oceanic islands, before human introductions, are wildly unrepresentative of those found elsewhere. For example, reptiles and birds are usually present, whereas terrestrial mammals and amphibians are consistently missing. In New Zealand, there were no terrestrial mammals before human occupation, though there were two species of bats. This reflects the ability of bats to cross large bodies of salt water. The rampant spread of many species after human introduction shows clearly that the local conditions were not unsuitable for their establishment. But even among the major types of animals and plants that are present, whole groups are often missing, whereas others are disproportionately represented. Thus, on the Galapagos islands, there are just over 20 species of land birds, of which 14 are finches, the famous finches described by Darwin in his account of his voyage round the world in *HMS Beagle*. This is quite unlike the situation elsewhere, in which finches form only a small fraction of the land bird fauna. It is exactly what one would expect if there were only a small number of species of original bird colonists, one of which was a species of finch that became the ancestor of the present-day species.

As this view would predict, oceanic islands provide many examples of forms that are unique to them, but nevertheless show affinities to mainland species. For example, 34% of the plant species found on the Galapagos islands are present nowhere else. Darwin's finches have a much greater variety of beak sizes and shapes than is usual among finches (which are normally seed-eaters with large, deep beaks), and these are clearly adapted to different modes of food gathering (Figure 12). Some of these are highly unusual, such as the habit of the sharp-beaked ground finch *Geospiza difficilis* of pecking the rear ends of nesting seabirds and drinking their blood. The woodpecker finch *Cactospiza pallida* uses twigs or cactus spines to extract insects from dead wood. Even more spectacular examples of rampant evolution are found on other groups of oceanic islands. For instance, the number of species of the fruitfly *Drosophila* on Hawaii exceeds that found in the rest of the world, and they are amazingly diverse in body size, wing patterns, and feeding habits.

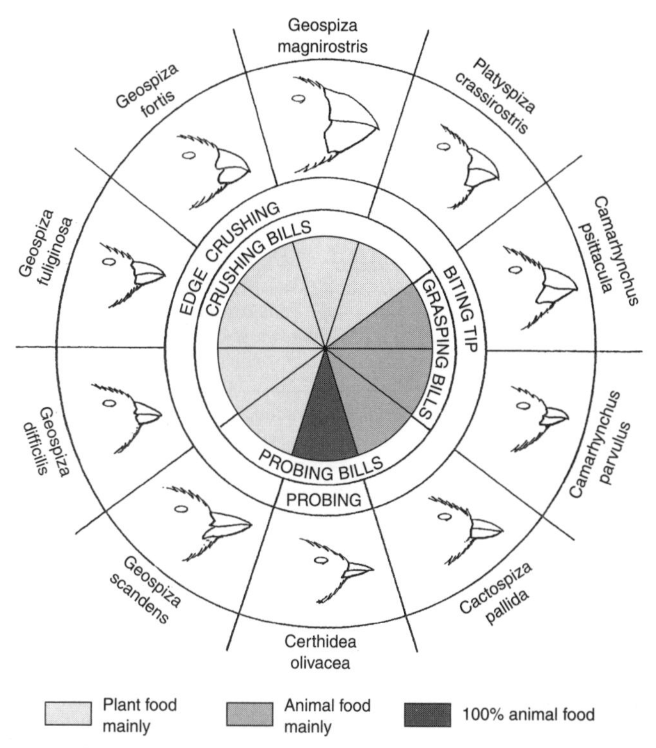

12. The beaks of Darwin's finches, showing the differences in size and shape between species with different diets.

These observations are explicable if the colonist ancestors of these island species found themselves in environments free from established competitor species. This situation would permit the evolution of traits that adapted the colonists to new ways of life, and allowed diversification of an ancestral species into several descendant species. Despite the unusual modifications of structure and behaviour found in Darwin's finches, studies of their DNA, by the methods described in Chapters 3 and 6, show that these species have a relatively recent origin about 2.3 million years ago, and are closely related to species on the mainland (Figure 13).

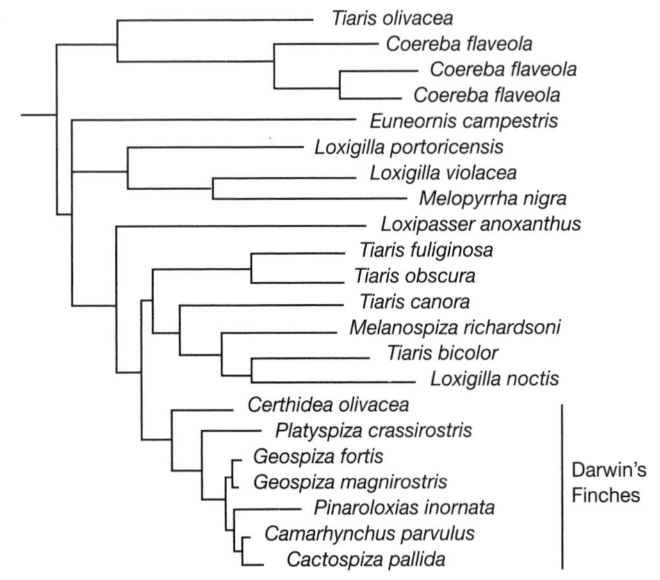

13. Phylogenetic tree of Darwin's finches and their relatives. The tree is based on differences among different species in the DNA sequences of a gene in their mitochondria. The lengths of the horizontal branches in the tree indicate the amounts of differences between species (ranging from 0.2% between the closest species to 16.5% between the most different). The tree shows that the Galapagos species form a cluster clearly having a common ancestor, and that they all have similar sequences of this gene, consistent with this ancestor being quite recent. In contrast, the other related species of finch differ much more from one another.

As Darwin wrote, when describing the inhabitants of the Galapagos islands in *The Origin of Species*:

> Here almost every product of the land and of the water bears the unmistakable stamp of the American continent. There are twenty-six landbirds; and twenty-five of these are ranked by Mr. Gould as distinct species, supposed to have been here created here; yet the close affinity of most of these birds to American species in every character, in their habits, gestures and tones of voice, was manifest.

> So it is with the other animals, and with nearly all the plants, as shown by Dr. Hooker in his admirable memoir on the Flora of this archipelago. The naturalist, looking at the inhabitants of these volcanic islands in the Pacific, distant several hundred miles from the continent, yet feels that he is standing on American land. Why should this be so? why should the species which are supposed to have been created in the Galapagos Archipelago, and nowhere else, bear so plainly a stamp of affinity to those created in America? There is nothing in the conditions of life, in the geological nature of the islands, in their height or climate, or in the proportions in which the different classes are associated together, which resembles closely the conditions of the South American coast; in fact there is a considerable dissimilarity in all these respects.

The theory of evolution, of course, provides the answer to these questions; research on island life over the past 150 years has amply confirmed Darwin's remarkable insights.

Chapter 5
Adaptation and natural selection

The problem of adaptation

An important task of the theory of evolution is to account for the diversity of living organisms within the hierarchical organization of resemblances between them. In Chapter 3, we emphasized resemblances between different groups, and how they make sense in terms of Darwin's theory of descent with modification. The second essential part of evolutionary theory is to provide a scientific explanation for the 'adaptation' of living organisms: their appearance of good engineering design, and their diversity in relation to their different ways of living. This will require the longest chapter in this book.

There are innumerable remarkable examples of adaptations, and we will just mention a few to illustrate the nature of the problem. The diversity of different kinds of eyes alone is astonishing, and yet makes sense in relation to the environments in which different animals live. Eyes for seeing underwater are different from those for seeing in air, and the eyes of predators have special adaptations to break the camouflage of prey that have evolved to be difficult to see. Many underwater predators that eat transparent marine animals have eyes with special contrast-increasing systems, including ultra-violet vision and polarized light vision. Other well-known adaptations are the hollow bones of birds' wings, with internal

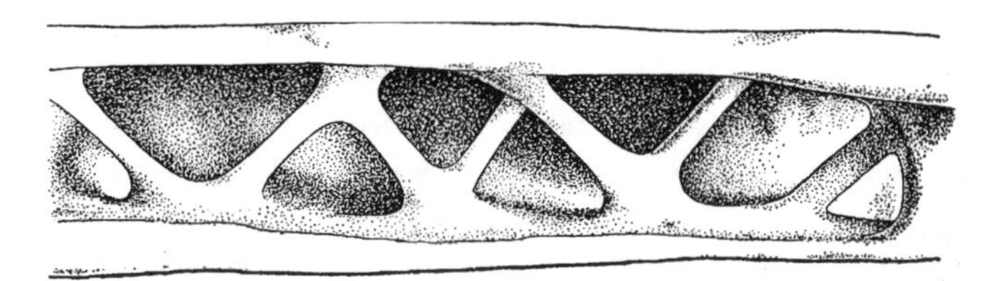

14. A hollow bone of a vulture's wings, with its internal strengthening struts.

struts resembling those in aircraft wings (Figure 14), or the wonderful construction of animal joints, whose surfaces allow the moving parts to move smoothly over one another.

Many other examples are provided by animal adaptations related to their different ways of feeding, and by reciprocal adaptations of the organisms that they feed on. Butterflies have long tongues to reach deep down into flowers and suck up nectar; reciprocally, flowers have high visibility to insects, and advertise themselves by scents, as well as rewarding visitors by nectar. Frogs and chamaeleons have long tongues that can shoot out and capture insect prey on their sticky tips. Many animals have adaptations to help them escape from predators, and the appearance of such animals depends on where they live. The silvery colour of many fish species makes them difficult to see in the water, but few land animals have this coloration. Some animals have cryptic coloration, with extraordinarily close mimicry of leaves or twigs, or of other poisonous or stinging species.

Adaptations are recognizable in many details of animals', plants', and microbes' lives, at every level, down to the cellular machinery and its controls (which we described in Chapter 3). For instance, cell division and cell movements are powered by tiny motors made of protein molecules. Proof-reading of newly produced DNA occurs when the genetic material is copied while making a new cell, reducing the frequency of harmful mutations several-thousand-fold. Protein complexes in cell surfaces selectively allow entry of some chemicals, but prevent others getting in. In nerve cells, these are used to control the flow of electrically charged metal atoms across the cell surface, generating the electrical signals used in transmitting information along the nerves. The behaviour patterns of animals are the ultimate outcome of the patterns of activity of their nerves, and are clearly adapted to their ways of life. For instance, in birds, nest parasites like cuckoos remove the host species' eggs or young from the nest, leaving the hosts to raise their young. In turn, the host species adapt by becoming more vigilant. Ants that grow fungus 'gardens' have evolved behaviours including weeding out spores of fungi contaminating their decaying leaves. Even the rate of ageing is adapted to the environment an animal or plant lives in, as we shall explain in Chapter 7.

Before Darwin and Wallace, such adaptations appeared to require a Creator. There seemed no other way to account for the astonishing detail and apparent perfection of many aspects of living organisms, just as the complexity of a watch could not be a purely natural production. The absence of any other explanation was the main support for the *Argument from Design* developed by 18th-century theologians to 'prove' the existence of a Creator, and the term *adaptation* was introduced to describe the observation that living things have structures that seem to be useful to them. It is important to understand that describing these as adaptations poses a question. It was a valuable contribution to our understanding of life to see that adaptations demand an explanation.

There is no doubt that animals and plants differ from other naturally produced things, such as rocks and minerals, as we acknowledge in the game 'animal, vegetable, or mineral'. But the Argument from Design overlooks the possibility that there could be natural processes, in addition to those that produce minerals and rocks, mountains and rivers, which can account for living creatures as complex natural productions, without the need for a Designer. The biological explanation of the origin of adaptation replaces the idea of a Designer, and is central to post-Darwinian evolutionary biology. In this chapter, we describe the modern theory of adaptation and its biological causes and basis. This is based on the theory of natural selection, which we outlined in Chapter 2.

Artificial selection and heritable variation

A first, very pertinent observation, strongly emphasized by Darwin, is that the modification of organisms by humans is regularly possible, and can produce the same appearance of design that we see in nature. This is routinely achieved by *artificial selection*, or selective breeding from animals and plants with desirable characters. Very striking changes can be produced over a time-frame that is short on the scale of the fossil record of evolution. For example, we have developed many different strains of cabbages, including strange ones like the cauliflower and broccoli, which are mutants that cause monstrous flowers forming a large head, and ones like the brussels sprout in which leaf development is abnormal (Figure 15A). Similarly, many breeds of dogs have been bred by humans (Figure 15B), with differences very like those observed between different species in nature, as Darwin pointed out. However, although all *Canis* species (including coyotes and jackals) are close relatives and can interbreed, dog breeds are not domestications of different wild dog species, but have been produced over the past few thousand years (several hundred dog generations) by artificial selection from a single common ancestral species, the wolf. The DNA sequences of dog genes are essentially a subset of wolf sequences, but coyotes (whose ancestor is believed from fossils to have separated from wolves' ancestors a million years ago) are about three times as different from either dogs or wolves as the most different dog/wolf comparison. The differences among dogs in their sequences of the same gene, differences which presumably developed after dogs separated from wolves, can be used to tell how long ago that separation happened (see Chapter 3). The conclusion is that dogs separated from wolves much longer ago than 14,000 years, the date suggested by archaeological records, but not more than 135,000 years ago.

A

Kale Brussels sprouts Broccoli Kohlrabi Cabbage Cauliflower

B

15. A. Some of the diversity of cultivated varieties of cabbage. B. Differences in the sizes and shapes of two breeds of dog.

The success of artificial selection is possible because heritable variation exists within populations and species (the slight differences between normal individuals which we described in Chapter 3). Even without any understanding of inheritance, people have bred from animals and plants that had characteristics they liked or found useful, and over enough generations this process has generated strains of animal and plant species that differ greatly from one another, and from the ancestral forms that were originally domesticated. This shows clearly that individuals within domesticated species must have been different from one another, and that many differences can be passed from parents to their offspring, that is they are heritable. If differences were merely due to the way the animals or plants were treated, selective breeding and artificial selection would have no effect on the next generation. Unless some of the differences were heritable, the breed could improve only by improved husbandry.

Every imaginable kind of character can vary heritably. The different breeds of dog differ, as is well known, not just in appearance and size, but also in mental traits such as character and disposition, some tending to be friendly, while others are fierce and suitable as guard dogs. They differ in their interest in scents, and in their inclination to fetch and carry or to swim, and in intelligence. They differ in the diseases to which they are susceptible, as in the well-known case of Dalmatians being prone to gout. They even differ in the ageing process, with some breeds, such as the Chihuahua, having surprising longevity (their life-span is almost as long as that of cats), while others, such as the Great Dane, live only about half as long. Although all these characteristics are, of course, affected by environmental circumstances such as good care and treatment, they are strongly influenced by heredity.

Similar heritable differences are known in many other domesticated species. To take another example, the qualities of different apple varieties are heritable differences. They include adaptations to different human needs such as early or late harvesting, suitability for cooking or eating, and to the differing climates of different countries. Just as in the case of dogs, other evolutionary processes have gone on in apples at the same time as human selection, and perfection is never attained for all desirable traits. For instance, Coxes are a particularly flavoursome apple, but are highly susceptible to disease.

Kinds of heritable variation

The success of artificial selection is very strong evidence that many kinds of character differences in animals and plants are heritable. There are also many genetic studies showing heritable variation for the characteristics of a wide range of organisms in nature, including many different species of animals, plants, fungi, bacteria, and viruses. Variation originates by well-understood processes of random mutation in the DNA sequences of genes, similar to those that produce human genetic disorders (Chapter 3). Most mutations are probably deleterious, like the genetic disorders of humans and farm animals, but advantageous mutations do sometimes occur. Such mutations have led to the resistance of animals to disease (such as the evolution of myxomatosis resistance in rabbits). They are also responsible for a major problem today, pests evolving resistance to chemicals (including resistance of rats to warfarin, or of worms parasitic in humans and farm animals to antihelminthic chemicals, insecticide resistance in mosquitoes, and antibiotic resistance in bacteria). Because of their importance to human or animal welfare, many cases are understood in great detail.

Heritable differences are also well known in humans. Variation may take the form of 'discrete' character differences, such as eye and hair colour, as already mentioned. These are variants controlled by differences in single genes, and unaffected by environmental circumstances (or altered only slightly, for instance when a fair-haired person's hair is bleached by the sun). Common variants like these are called *polymorphisms*. Conditions such as colour blindness are also simple genetic differences, but are much rarer variants in human populations. Even behavioural characters may be heritable. Whether fire ant colonies have single or multiple queens seems to be controlled by a difference in a single gene for a protein that binds a chemical involved in recognition of other individuals.

'Continuous' variation is also very evident for many characters in populations, for example the gradations of height and weight among people. This kind of variation is often markedly affected by environmental conditions. The increasing height of successive generations during the 20th century, seen in many different countries, is not due to genetic changes but to changed conditions of life, including better nutrition and fewer serious illnesses during childhood. Nevertheless, there is also some degree of genetic determination for such characters in human populations. This is known from studies of identical and non-identical twins. Non-identical twins are ordinary siblings that happen to be conceived at the same time, and they differ as much as any siblings, but identical twins come from a single fertilized egg that splits into two embryos, and are genetically identical. Greater resemblances between identical than non-identical twins have been documented for many characteristics, which must be due to their genetic similarity (care must, of course, be taken that the identical twins are not treated more alike than non-identical pairs – for instance, only same-sex pairs of both kinds of twins should be studied). Despite the important environmental influences that clearly often exist, this and other kinds of evidence consistently reveal some degree of heritable basis for variation in many characteristics, including aspects of intelligence. Heritable variation has been documented in many organisms, and for all kinds of characteristics. Even an animal's place in the dominance hierarchy, or pecking order, can be heritable; this has been demonstrated in chickens and in cockroaches. The amount of continuous genetic variability can be measured from resemblances between relatives of different degrees. This is useful in animal and crop plant breeding, and allows breeders to predict the characteristics, such as milk yield of cows, that offspring of different parents will have, and thus to plan their breeding.

Genetic differences boil down to differences in the 'letters' in the DNA. These often leave the amino acid sequences of proteins unchanged. When the DNA sequences of the same gene are compared between different individuals, differences are seen, though usually fewer than when sequences are compared between different species (such comparisons were discussed in Chapter 3, see Figure 8). For example, copies of the gene for glucose-6-phosphate dehydrogenase, mentioned in Chapter 3, one from each of a set of individuals, might be compared. There may be no differences (so no diversity). If some individuals in the population have a variant sequence of the gene, the difference will show up in some of the comparisons. This is called molecular polymorphism. Geneticists measure such diversity by the fraction of the letters in the DNA sequence that vary between individuals in the population. In the human species, it is usually found that fewer than 0.1% of the DNA letters differ when we compare the same gene's sequence between different people (compared with generally around 1% of the letters being different when a gene's sequences are compared between a human and a chimpanzee). Variation is higher in some genes and lower in others, and, as one might expect, variation is generally higher in the presumably less important regions of the genome that do not code for proteins than in the coding parts of genes. Humans are rather lacking in variability compared with most other species. For example, DNA polymorphism is much more common in maize (more than 2% of its DNA letters are variable).

The distribution of variability within a species can give us useful information. When dogs are bred for different characteristics, breeds are developed that are rather uniform in their characteristics. This is due to strict pedigree rules, which control matings and forbid 'gene flow' between breeds. A characteristic that is desired in one breed, such as fetching, is thus well developed in that breed only, and separate breeds tend to diverge from each other. This isolation between breeds is unnatural, and dogs of different breeds will happily mate and produce healthy young. Much of the variability of dogs is accordingly between breeds. Many natural species live in different, geographically separated populations, and, as one might expect, the amount of diversity in such species as a whole is greater than within a single population, because there are differences between populations. For example, certain blood groups are more common in some human races than in others (see Chapter 6), and the same is true for many other genetic variants. However, in humans and many other species in nature, the differences between populations are very slight in comparison with the diversity within populations, unlike the situation for dog breeds. The difference is because humans move freely between populations. An important implication of these genetic results is that human races are distinguished by a small minority of the genes in our genomes, most of our genetic make-up having a similar range and heterogeneity of variants worldwide. Increased mobility in the modern world is quickly reducing any differences between populations.

Natural selection and fitness

A fundamental idea in the theory of evolution under natural conditions is that some heritable character differences affect survival and reproduction. For instance, just as race horses have been selected for speed (by breeding from winners and their relatives), so antelopes have been naturally selected for speed, because the individuals that breed and contribute to the future of their species are those that did not get eaten by predators. Darwin and Wallace realized that this kind of process could explain adaptation to natural conditions. Our ability to modify animals and plants by artificial selection depends on this characteristic having a heritable basis. Provided that there are heritable differences, successful individuals in the wild will likewise pass their genes (and thus often their good characteristics) to their offspring, which will, in turn, possess the adaptive characters, such as speed.

For brevity, and to allow one to think in general terms, the word *fitness* is often used in biological writing to stand for overall ability to survive and reproduce, without the need to specify which characters are involved (just as we use the term 'intelligence' to mean a variety of different abilities). Many different aspects of organisms contribute to fitness. For instance, speed is just one feature affecting antelope fitness. Alertness and the ability to detect predators are also important. Mere survival is not enough, however, and reproductive abilities, such as provisioning and care of the young, are also important for fitness in animals, and the ability to attract pollinators is critical for fitness in flowering plants. The word fitness can accordingly be used to describe selection acting on a wide range of different traits. As with 'intelligence', the generality of the term 'fitness' has led to misunderstandings and disputes.

To know what characters are likely to be important for the fitness of an organism, one must understand a great deal about its biology and the environment in which it lives. The same character may give high fitness in one species, but not in another. For instance, speed is not important for fitness in a lizard that evades predators by cryptic coloration. If such a lizard lives in trees and perches on twigs, it is more important for it to be good at holding on than to run fast, and so short legs, not long ones, will be associated with high fitness. Speed is adaptive for antelopes, but staying very still, so as not to be detected by predators, is an alternative means by which many animals avoid being eaten. Other animals avoid predators by frightening them away; for example, some butterflies have eye spots in their wing patterns that can be suddenly displayed in order to alarm birds. Plants obviously cannot move, and avoid being eaten by different means, including tasting bad or being prickly. All these different characteristics may increase the survival and/or reproduction of the organisms, and hence their fitness.

Given genetic variability for many characters, and environmental differences, natural selection will inevitably operate, and the genetic make-up of populations and species will change over time, as we showed in Chapter 2. Changes are often slow in terms of years, because it takes many generations for a genetic variant that is rare to become the majority type in the population. In animal and plant breeding, severe selection often occurs (for example, when diseases wipe out most of a herd or a plant crop), but changes still take many years. It is estimated that maize was domesticated about 10,000 years ago, but modern giant corn cobs are a quite recent development. Despite the slowness of evolutionary change in terms of years, natural selection can produce rapid changes on the timescale of the fossil record. Advantageous traits can spread throughout a population from a very low starting frequency in less time than that between successive layers in the geological strata (usually at least several thousand years, see Chapter 4).

Even though we often may not see it happening, because of its slowness in terms of the time-scale of our lives, natural selection never stops. Even humans are still evolving. For instance, our diet differs from that of our ancestors, and our teeth can function quite well on soft modern foods even if they are not very strong. The high sugar content of many modern foods leads to tooth decay, and potentially to abscesses that can be fatal, but there is no longer very pronounced natural selection for strong teeth, because dental care can solve these problems, or provide false teeth. Just as for other functions that are no longer used intensively, changes are to be expected, and our teeth could one day become vestigial. They are already smaller than those of our close relatives, the chimpanzees, and there is no reason why they should not become smaller still. Excess sugar in the diet has also led to an increasing frequency of late-onset diabetes in human populations, with high mortality for sufferers. In the past, this disease was largely confined to people past childbearing age, but the age of onset is becoming steadily earlier. There is therefore a new, probably intense, selection pressure to change our metabolism so as to tolerate our changed diet. In Chapter 7, we will show how changes in human life are leading to the evolution of greater longevity.

The concept of fitness is often misunderstood. When biologists try to illustrate the meanings of this term, they often use examples that correspond with our everyday use of the word fitness, such as the speed of antelopes. There is less danger of confusion if we think of characteristics like the lightweight bones of birds, with their hollow centres and strengthening cross-struts (Figure 14). The theory of natural selection accounts for such apparently well-designed structures by pointing out that, when flight was evolving, lighter-boned individuals would have had slightly higher chances of survival than others. If their descendants inherited lighter bones, the characteristic would increase in its representation in the population over the generations. This is just the same as artificial selection by breeders of the fastest dogs, which has given all greyhounds long, thin legs. These are mechanically more efficient than short ones, and greyhounds' legs closely resemble those of antelopes and other fast-running animals, which have evolved by natural selection. We can describe natural and artificial selection perfectly well without using the word fitness. Natural selection implies nothing more than that certain heritable variants may be preferentially passed on to future generations. Individuals carrying genes that lower their survival or reproductive success will generally not pass on those genes to the same extent as other individuals whose genes give higher survival or reproductive ability. The term fitness is merely a useful short-cut to help express briefly the idea that characteristics sometimes affect organisms' chances of surviving and/or reproducing, without having to specify a particular characteristic. It is also useful in making mathematical models of the way selection affects the genetic make-up of a population. Conclusions from these models provide a rigorous underpinning for many of the statements that we make in this chapter, but we will not describe them here.

To illustrate selection of an advantageous mutation, consider the arms race between humans and rats, in which we try to develop rat poisons, and rats evolve resistance. The rat poison warfarin kills rats because it prevents blood clotting. It binds to an enzyme needed in the metabolism of vitamin K, which is important for blood clotting and many other functions. Resistant rats were once rare, because their vitamin K metabolism is changed, reducing growth and survival. In other words, there is a *cost* of resistance. In farms and towns where warfarin is used, however, only resistant animals can survive, so there is strong natural selection, despite the cost. The resistant version of the gene has therefore spread to high frequencies in the rat population, though the cost keeps it from spreading to all members of the species. However, a recent development is the evolution of a new kind of resistance which seems to be free from the cost, and which may even be advantageous (in the absence of poison). There is thus continued evolution in response to a change in the rats' environment.

Variability and selection are very general properties of many systems, not just individual organisms. Certain components of the genetic material are maintained, not because they increase the fitness of the organisms that carry them, but because they can multiply within the genetic material itself, just like parasites in the body of their host. 50% of human DNA is thought to belong in this category. Another important situation in which natural selection drives evolutionary change within an organism occurs in cancers. Cancer is a disease in which a cell and its descendants evolve selfish behaviour and multiply, regardless of the good of the rest of the body. The disease is often caused by a mutation that increases the mutation rates of other genes (for instance, by a failure in the proof-reading system described in Chapter 3, which checks DNA sequences and prevents mutations). If mutations occur at a high frequency, some may affect cell multiplication rates, and a fast-multiplying lineage may appear. As time goes on, more and more of the cells will descend from cells carrying mutations in other genes which confer faster and faster growth, and so the cancer often becomes more aggressive. Cancer cells can also become resistant to drugs that suppress their growth. Like the well-known situation of drug-resistant HIV viruses evolving in an AIDS patient, cancer cells which acquire mutations that allow them to escape drug suppression outgrow the initial type of cells, and cause loss of remission of cancers. This is why it is often hopeless to restart drug treatment after a remission stops.

At the other extreme, there may be different rates of extinction of species with different sets of characteristics, that is there can be selection at the level of species. For example, species with large body sizes, which tend to have low population sizes and low rates of reproduction, are more vulnerable to extinction than species with small bodies (see Chapter 4). In contrast, selection between individuals of the same species often favours larger body size, probably because larger individuals have greater success in competition for food or mates. The range of body sizes that we see in a group of related species may reflect the net outcome of both types of selection. Selection on individuals within species is likely, however, to be the most important factor, since it produces the different range of body sizes in the first place, and it usually operates much faster than selection at the species level.

Selection is also important in non-biological contexts. In designing machines and computer programs, it has been found that a very efficient way to find the optimal design is to successively make small, random changes to the design, keeping versions that do the job well, and discarding others. This is increasingly being used to solve difficult design problems for complex systems. In this process, the engineer does not have a design in mind, but only the desired function.

Adaptations and evolutionary history

The theory of evolution by natural selection explains features of organisms as a result of the successive accumulation of changes, each giving higher survival or reproductive success. What changes are possible depends on the pre-existing state of the organism: mutations can only modify the development of an animal or plant within certain limits, which are constrained by the underlying existing developmental programmes that lead to the adult organism. The results of artificial selection as practised by animal and plant breeders show that it is relatively easy to change the sizes and shapes of body parts, or to produce striking changes in superficial characters such as external coloration, as in different breeds of dog. Radical changes can easily be produced by mutations, and laboratory geneticists have no difficulty in creating strains of mice or fruitflies that differ much more from normal forms than wild species differ from each other. It is possible, for example, to produce flies with four wings instead of the normal two. These major changes, however, often severely disrupt normal development, reducing survival and fertility, and are therefore unlikely to be favoured by natural selection. They even tend to be avoided by animal and plant breeders (although such mutations have been used in developing unusual pigeon and dog breeds, where the animals' health is of lesser importance than for farmers).

For this reason, we expect that evolution will usually proceed by fairly small adjustments to what has gone before, rather than by sudden jumps to radically new states. This is particularly obvious for complex traits that depend on the mutual adjustment of many different components, such as the eye (which we discuss in more detail in Chapter 7); if one component is changed drastically, it may not function well in combination with other parts that remain unchanged. When new adaptations evolve, they will usually be modified versions of pre-existing structures, and will at first often not be the optimum functional engineering design solutions. Natural selection resembles an engineer improving machinery by tinkering with it and modifying it, rather than sitting down and planning entirely new designs. Modern screwdrivers can be suitable for precision work, with a diversity of heads suited for different purposes, but the evolutionary ancestors of screws were coarse-threaded spigots turned by a spike through a hole in one end.

While we are often astonished by the precision and efficiency of adaptations of living organisms, there are many examples of

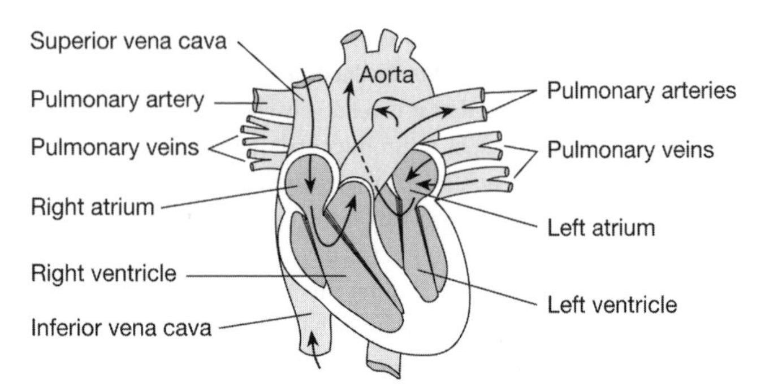

16. The highly complex structure of the mammalian heart and its blood vessels. Note how the pulmonary artery (which delivers blood to the lungs) curves awkwardly back behind the aorta (which delivers blood to the rest of the body) and the superior vena cava (which brings blood back to the heart from the head).

tinkering, betrayed by features that make sense only in terms of their historical origins. Painters represent angels with wings on their shoulders, allowing them the continued use of their arms. But the wings of all real flying or gliding species of vertebrates are modified forelimbs, so that pterodactyls, birds, and bats have all lost the use of their forelimbs for most of their original functions. Similarly, the design of the mammal heart and circulation has bizarre features that reflect a history of gradual modification from a system that originally functioned to pump blood from the heart round the gills of a fish, and then to the rest of the body (Figure 16). The embryonic development of the circulatory system clearly betrays its evolutionary antecedents.

Sometimes, similar solutions to a functional problem have been evolved independently in different groups, resulting in very similar adaptations that nevertheless differ considerably in detail, because of their different histories, as in the case of the wings of birds and bats. Thus, while the similarity of different organisms is often due to their being related (like ourselves and the apes), distantly related organisms living in similar circumstances can sometimes look more similar than closer ones. When morphological similarities and differences are misleading, the true evolutionary relationships can be discovered using evidence from DNA sequence similarities and differences, as explained in Chapter 3. For example, several species of river dolphins have evolved in great rivers in different parts of the world. They share some features differentiating them from open-sea species, particularly reduced eyes, because they live in turbid water, and rely more on echo-location than vision for navigation. DNA sequence comparisons show that a given species of river dolphin is more closely related to the sea-living species in its region than to river dolphins elsewhere. It makes sense that similar environments lead to similar adaptations.

Despite the similarities, natural selection differs from human design processes in several ways. One difference is that evolution has no foresight; organisms evolve in response to prevailing environmental conditions at one time, and this may result in features which lead to their extinction when conditions change radically. As we show later in this chapter, sexual competition among males can lead to structures that severely reduce their survival ability; it is quite possible that in some cases an unfavourable environmental change could further reduce survival to such a point that the species could not maintain itself, as has been suggested for the extinct Irish Elk, with its enormous antlers. Long-lived organisms often evolve very low fertility, as in the case of large birds of prey such as condors that only produce one offspring every other year (we discuss this further in Chapter 7). Such populations can do well if the environment is favourable, and there is low annual mortality of breeding adults. However, if the environment deteriorates and mortality increases, for example because of human disturbances, this may cause a rapid decline in population number. This is happening at the present time to many species, and has caused the extinction even of species that were once very abundant. For instance, the slow-breeding passenger pigeon of the USA was hunted to extinction in the 19th century despite originally having populations of tens of millions. Species which evolve to occupy an extremely specialized type of habitat are also vulnerable to extinction if that habitat disappears due to environmental change; for example, pandas in China are under threat because they breed slowly and depend on a type of bamboo found only in certain mountainous regions, which are now being logged.

Natural selection also does not necessarily produce perfect adaptation. In the first place, there may not be time to adjust every aspect of a piece of biological machinery to its best-functioning state. This is particularly likely to be true when selection pressures result from interactions between a pair of species, such as a host and a parasite. For example, an improvement in the ability of the host to resist infection increases the pressure of selection on the parasite to overcome this resistance, forcing the host to evolve new resistance measures, and so on, so that is there is an 'evolutionary arms race'. In such situations, neither partner can remain perfectly adapted for very long. Despite the wonderful ability of our immune system to combat bacterial and viral infections, we remain vulnerable to newly evolved strains of influenza and cold viruses. Second, the tinkering aspect of selection, modifying what has gone before, constrains what selection can achieve, as we have just mentioned. It seems absurd from a design point of view that the nerves that carry information from the light-sensitive cells of the vertebrate eye are in front of, rather than behind, the light-sensitive retinal cells, but this is a consequence of the way this part of the eye develops as an outgrowth of the central nervous system (the octopus eye resembles that of mammals, but has a better arrangement, with the light-sensitive cells in front of the nerves). Third, an improvement in one aspect of the functioning of a system may have a cost with respect to some other function, as mentioned in relation to warfarin resistance. This can prevent improved adaptation. We will mention other examples later in this chapter, and in Chapter 7, when we discuss ageing.

Detecting natural selection

Darwin and Wallace argued that natural selection is the cause of adaptive evolution without knowing examples of selection operating in nature. Over the last 50 years, many cases of natural selection have been detected in action and studied in detail, immeasurably strengthening the evidence for its key role in evolution. We have space for only a few examples. A very important kind of natural selection acting today is causing ever-increasing antibiotic resistance in bacteria. This is an example of evolutionary change that is intensively studied, because it endangers our lives, and occurs fast and (unfortunately) very repeatedly. On the day we were writing this, the headlines in the newspaper were about methicillin-resistant *Staphylococcus* in Edinburgh's Royal Infirmary. Whenever an antibiotic is widely used, resistant bacteria are soon found. Antibiotics were first widely used in the 1940s, and concerns about resistance were soon being raised by microbiologists. In 1955, an article in the *American Journal of Medicine*, aimed at doctors, wrote that the indiscriminate use of antibiotics: 'is fraught with the risk of selecting resistant strains', and in 1966 (when people had not changed their behaviour), another microbiologist wrote: 'is there no way to generate sufficient general concern so that antibiotic resistance can be attacked?'

The speedy evolution of antibiotic resistance is not surprising, because bacteria multiply fast and are present in enormous numbers, so that any mutation that can make a cell resistant is sure to occur in a few bacteria in a population; if the bacteria are able to survive the change to their cell functions caused by the mutation and to multiply, a resistant population can rapidly build up. One might hope that resistance will be costly for bacteria, as was initially true for warfarin resistance in rats, but as in rats we cannot rely on this remaining true for long. Sooner or later, bacteria will evolve so that they survive well in the presence of antibiotics, without serious costs to themselves. Our only chance is therefore to use antibiotics sparingly, confining use to situations where they are really needed, and making sure that all infecting bacteria are killed quickly, before they have time to evolve resistance. If one stops treatment while some bacteria remain present, their population will inevitably include some resistant bacteria, which can then spread to other people. Antibiotic resistance can also spread between bacteria, even ones of different species. Antibiotics given to farm animals, to keep infections down and promote growth, can cause resistance to spread to human pathogens. Even these consequences are not the whole of the problem. Bacteria that have resistance mutations are not typical of their populations, but sometimes have higher mutation rates than average, allowing them to respond even faster to selection.

Drug and pesticide resistance evolve whenever drugs are used to kill parasites or pests, and literally hundreds of cases have been studied in microbes, plants, and animals. Even the HIV virus mutates within AIDS patients treated with drugs, and evolves resistance so that the treatment eventually fails. To try to prevent this, two drugs instead of one are often used. Because mutations are rare events, the virus population in a patient is unlikely to get both resistance mutations very quickly, but this usually happens eventually.

These examples illustrate natural selection, but, like artificial selection, they involve situations where the environment is changing as a result of human intervention. Many other human activities are causing evolutionary changes in organisms. For example, it seems that killing elephants for their ivory has led to increased frequencies of tuskless elephants. In the past, these were rare, abnormal animals. Now, with intensive hunting, these animals can survive and reproduce better than normal ones, and as a consequence they are increasing in elephant populations. Swallowtail butterflies with small wings, which are poor flyers, are being selected in fragmented natural habitats, presumably because individuals that do not fly far are more likely to remain within suitable habitat patches. Humans also select for an annual life-history, with speedy seed production, when we remove weeds from gardens or crop fields. In species such as the grass *Poa annua*, individuals exist that develop more slowly, and can live two years or more, but these are at a clear disadvantage in a regime of intensive weeding. These examples not only show how common and rapid evolutionary change can be, but also that anything we do may affect the evolution of species associated with humans. With people spreading all over the planet, few species will remain unaffected.

Biologists have also studied many cases of selection that are entirely natural, not involving human habitat degradation or alteration. One of the best is the 30-year study by Peter and Rosemary Grant of two species of Darwin's finch, the ground finch and the cactus finch, on the island of Daphne in the Galapagos islands (see Chapter 4). These species differ in their mean beak sizes and shapes, but there is considerable variation within each species for both characters. During the study, the Grants' team systematically ringed and measured every bird hatched on the island, and the offspring of every female were identified. Their survival through life was followed and related to measurements of size and shape of body parts. Pedigree studies showed that the variation in beak characters has a strong genetic component, so that offspring resemble their parents. Studies of the feeding behaviour of the birds in the wild show that beak size and shape affect the efficiency with which the birds deal with different types of seed: large, deep beaks allow birds to handle large seeds better than small, shallow beaks, while the reverse is true for small seeds. The Galapagos are subject to episodes of severe drought, associated with the El Niño phenomenon, and these affect the abundances of different types of food. In a drought year, most food plants fail to produce seeds, except for a species that produces very large seeds. This means that birds with large, deep beaks have a much better chance of survival than others, as was seen directly from the population censuses: after an episode of drought, the surviving adults in both species had larger, deeper beaks than the population before the drought. In addition, their offspring inherited these characters, so that the change in direction of selection caused by the drought induced a genetic change in the composition of the population – a real evolutionary change. The extent of this change agreed with that predicted from the observed relation between mortality and beak characters, taking into account the degree of resemblance between parents and offspring. Once normal conditions were restored, the relations between beak characters and survival changed in such a way that large, deep beaks were no longer favoured, and the populations evolved back towards the previous state. However, even in non-drought years there were also more minor changes in the environment that resulted in changes in the relation between fitness and beak traits, so there were constant fluctuations in beak characteristics over the whole 30 years, with the populations of both species ending up significantly different from the initial state.

Another good example is provided by the way in which flowers are adapted to their insect and other animal pollinators. For a plant to mate with others of its own species, pollinators must be attracted to visit the plant's flowers, and be rewarded for doing so (by nectar or excess pollen that they can eat), which ensures that they will visit other plants of the same species. Both the plant and animal players in these interactions evolve to get the best they can for themselves. For an orchid, for instance, it is important that a pollinating moth probes the flowers deeply, in order that the pollen mass (called a pollinium) gets firmly attached to the moth's head when it visits. This ensures that the pollinium makes good contact with the right part of the flower which the moth visits next, so that it engages correctly and the pollen will fertilize the flower. The need to keep the nectar almost out of reach of the moths' tongues drives natural selection on nectary tube length, and flowers with deviant lengths of nectary tubes should therefore have lower fertility. Flowers with shorter tubes will allow moths to suck nectar without picking up pollinia or depositing them, and flowers whose tubes are too long will waste nectar, like juice boxes whose straws are invariably too short to get all the juice out. In the juice box industry, waste benefits juice sellers, enabling them to sell larger amounts, but plants lose energy, water, and nutrients if they make nectar that is useless, and these resources could be put to better use.

In a South African *Gladiolus* species that produces only one flower per plant, individuals with longer tubes produced a fruit more often than those with average tubes, and also had more seeds per fruit than the average. This species' tubes are on average 9.3 cm long, and their hawkmoth visitors' tongues are between 3.5 to 13 cm. Moths that had no pollen on their tongue all had the longest tongues. Other hawkmoth species at the same locality, which do not pollinate this species, have tongues averaging less than 4.5 cm. This shows the power of selection to push the flowers and moths to adapt to one another, reaching extremes in some cases. Some Madagascan orchids have flowers whose nectaries are 30 cm long, and their pollinators' tongues are 25 cm. In these species, selection on length has been demonstrated by experiments in which nectar spurs were tied off to shorten them, leading to lower success in getting the moths to remove pollinia.

A similar kind of selection and counter-selection affects our own species in its relations with parasites. Several different human adaptations to malaria have been well studied, and we have clearly evolved a number of different defences, including changes in our red blood cells, where malaria parasites spend part of their complex life-cycle. Like warfarin resistance in rats, defences may sometimes have costs. The disease sickle-cell anaemia, which is usually fatal in the absence of medical treatment, involves a changed haemoglobin (the major red blood cell protein, responsible for carrying oxygen round the body). The changed form (haemoglobin S) is a variant form of the gene that codes for the common adult haemoglobin, A, and the two versions differ by a single DNA letter. Individuals whose genes for this protein are both the S type suffer from sickle-cell anaemia; their red blood cells become malformed and clog tiny blood vessels. People with one normal haemoglobin A and one S version of the gene are not affected, but have the benefit of higher resistance to malaria compared with people with two haemoglobin

A genes. The disease suffered by people with two S genes is thus a cost of malaria resistance, and prevents the S form from spreading throughout the population, even in areas with high levels of malarial infections. The variants of the enzyme glucose-6-phosphate dehydrogenase that also help protect against malaria (see Chapter 3) have a cost as well, at least when people with these variants eat certain foods or drugs, causing damage to their red blood cells, whereas the non-resistant version of this enzyme prevents this. Resistance to malaria with no, or very slight, costs does seem to be possible, however. The blood type Duffy-, another red blood cell characteristic, is widespread in much of Africa, and people with this group are much less susceptible to a certain type of malaria than people who carry the alternative Duffy+ type.

Malaria resistance illustrates a common finding, that different responses can occur to a single selective pressure, in this case a serious disease. Some of the solutions to the problem posed by malaria are better than others, because there are lower costs for the people involved. In fact, there are many other genetic variants found in different human populations that confer resistance to malaria, and it seems to be largely a matter of chance which particular type of mutation becomes established by selection in a given locality.

The examples just discussed illustrate selection responses to changes in the environment of animals, humans, and plants. Perhaps a disease appears, and there is selection on a population so that resistant individuals evolve. Or perhaps a moth evolves a longer tongue and can take nectar from flowers without picking up pollen, so that the flower in turn evolves longer nectaries. In these examples, natural selection changes the organisms, as Darwin envisaged in the 1858 quotation we gave in Chapter 2. Natural selection, however, also often acts to prevent changes. In Chapter 3, when we described the cellular machinery of proteins and enzymes, we mentioned that mutations happen and can degrade these functions. Even in a constant environment, selection acts on every generation against mutant genes (that encode mutant proteins or ones that are expressed in the wrong place or time, or in the wrong amount). New individuals with mutations arise in every generation, but non-mutants tend to leave more offspring and so their genes remain the most common, with the mutated versions remaining at low frequencies in the population. This is *stabilizing* or *purifying* selection, keeping everything working as well as possible. An example is the gene that codes for one of the proteins involved in blood clotting. Some changes to the sequence of the protein result in an inability of blood to clot following a cut (haemophilia). Until quite recently, when the causes of haemophilia were understood and it became possible to help haemophiliacs by injecting clotting factor proteins, this condition was usually lethal or severely reduced survival. Thousands of low-frequency genetic variants with such deleterious effects, affecting every conceivable characteristic, have been described by medical geneticists.

Stabilizing selection occurs if the environment has remained fairly constant, so that selection in the past has had time to adjust a trait to the state which confers high fitness. It can be detected acting today on continuously variable characters of organisms. A well-studied example is human birth weight. Even today, when very few babies die, babies with intermediate weights survive best. The low level of infant mortality largely involves tiny babies, and some very large ones. Stabilizing selection has also been observed in animal species, such as birds and insects, after severe storms, when surviving individuals tend to be intermediate in size, while the smallest and largest are often lost. Even minor deviations from an optimum may lower survival or fertility. It makes sense, therefore, that organisms' adaptation to their environment is often impressive. As we explained in Chapter 3, it sometimes seems as if even the tiniest detail can be important. Near perfection is often achieved, such as the extraordinary accuracy of cryptic butterflies' resemblance to leaves or of caterpillars to twigs. Stabilizing selection also makes sense of the observation that species often show little evolutionary change; as long as their environment poses no new challenges, selection will tend to keep things as they are. The stable morphology of some organisms over long evolutionary times, such as the so-called *living fossils* whose modern members resemble distantly related fossils, can thus be understood.

Sexual selection

Natural selection is the only explanation for adaptation that has stood up to empirical tests. However, selection does not necessarily increase the overall survival or number of offspring produced by the population as a whole. When there is competition, traits that confer success in competition for a limited resource may reduce everyone's survival. If the most successful type of individuals become common in a population, the population's survival probability may be decreased. Examples of the maladaptive consequences of competition are not confined to biological situations. The intrusiveness and frequent bad taste of advertising are well known.

One of the best understood biological examples of competition is selection acting on the ability of males to obtain mates. In many animal species, not all fertile males leave descendants, but only those that succeed in courtship and/or in contests with other males. Sometimes, only the 'dominant' males are accepted by the females. Even fruitfly males have to court females – with dances, songs (produced by beating their wings), and scents – before being allowed to mate. Success is not guaranteed (not surprisingly, since females must be choosy and avoid accepting males of the wrong species). In many mammals, such as lions, there are social hierarchies in the ability to get matings, and females are choosy, so that males differ in their reproductive success. Natural selection will therefore favour characteristics associated with males' dominance in the hierarchy, or their attractiveness to females. Male deer have large antlers, which are used in fights between males, and some species have other means of intimidation, such as loud roaring. If

17. The outcome of sexual selection, as illustrated in Darwin's *The Descent of Man and Selection in Relation to Sex*. The figure shows a male and female of the same species of bird of paradise, showing the male's ornamentation and the female's lack of display.

these characters are heritable (which, as we saw above, is often the case), males with the characteristics that make them successful in mating will pass on their genes to many progeny, while other males will tend to have fewer offspring.

Both sexes may evolve characteristics by this *sexual selection*, and it probably accounts for the bright plumage of many birds. However, in many species these characteristics are confined to males (Figure 17), suggesting that they are not in themselves good adaptations to the species' environment. Many such male characters certainly do not seem likely to help survival, and they often incur costs because of lower survival of their male carriers. Peacock males, with their enormous and beautiful tails, are poor flyers, and they would probably be better able to escape predators if their tails were smaller. Peacocks are an inconvenient species for experimental studies of the aerodynamics of flight, but swallows' tails have been shown to be longer than optimal for flight, while males with longer tails are preferred by females. Even less spectacular male courtship characters often bring increased risks. For example, some tropical frog species are preyed on by bats that detect males singing their courtship songs. Even without these dangers, courting males often expend large amounts of effort, which could be otherwise employed, for example in looking for food, and they are often in extremely poor physical shape at the end of the mating season.

Realizing this, Darwin considered selection in the context of courtship to be different from most other situations, and introduced the special term sexual selection to highlight this difference. As we have just argued, it is unlikely that male peacocks' tails are adaptive, both on *a priori* grounds (such tails do not look like a good design for a flying animal), and because, if they *were* good, females should have them too. It therefore seems that selection has traded reduced flying ability against increased male mating successes in peacocks, a species in which competitive mating is important. Thus sexual selection again shows that the word fitness as used in biology often means something different from the everyday use of the word. A peacock male handicapped by his tail is not 'fit' in the meaning of being a good flyer or runner (although he may be unable to produce a fine tail if he is not well nourished and healthy), but in the shorthand of evolutionary biology he has high fitness; without his large tail, the females would mate with other males and his fertility would be low.

Chapter 6
The formation and divergence of species

One of the most familiar facts of biology is the division of living forms into recognizably different species. Even the most casual observation of the birds living in a north-western European town, for example, shows the presence of several species: the robin, blackbird, song thrush, missel thrush, blue tit, great tit, pigeon, sparrow, chaffinch, starling, and so on. Each species has its distinctive body size and shape, plumage coloration, song, and feeding and nesting habits. A different but broadly similar array of bird species can be found in eastern North America. Males and females of each species pair only with each other, and their offspring of course belong to the same species as their parents. Within a given geographical location, sexually reproducing animals and plants can nearly always readily be assigned to distinct groups (although careful observation sometimes reveals the existence of species with only very slight anatomical differences). Different species that coexist in the same locality remain distinct because they do not interbreed. Most biologists regard this lack of interbreeding (*reproductive isolation*) as the best criterion for defining different species. The situation is more complex with organisms that do not reproduce regularly by sexual matings, such as many kinds of microbes, and we will defer discussion of these until later.

The nature of differences between species

Although, like the force of gravity, this division of living organisms into discrete species is so familiar that we take it for granted, it is not an obviously necessary state of affairs. It is easy to imagine a world without such sharp differences; in the bird example above, there could be creatures that combine the characteristics of, say, robins and thrushes in different proportions, and in which a mating between a given pair of parents would yield offspring with widely different character combinations. If there were no barriers to interbreeding between members of different species, the diversity of life that we see in the world could not exist, and there would be something approaching a continuum of forms. In fact, when for one reason or another barriers to interbreeding between formerly separate species have broken down, such highly variable offspring are indeed produced.

A fundamental problem for evolutionists is therefore to explain how species come to be distinct, and why reproductive isolation exists. This is the main topic of this chapter. Before embarking on it, we will describe some of the ways in which closely related species are prevented from interbreeding. Sometimes, the main barrier is a simple difference in habitats or in the time of breeding of the species. In plants, for example, there is often a characteristic brief flowering time each year, and species with non-overlapping flowering times will obviously be unable to interbreed. In animals, the use of different breeding sites may prevent individuals from different species from mating with each other. Subtle features of organisms, which can only be discovered by detailed studies of the species' natural history, often prevent individuals from different species from successfully mating with each other, even if they come together in the same place at the same time. For example, there may be an unwillingness to court individuals of the other species, because they do not produce the right smell or sound, or their courtship displays may differ. Behavioural barriers to mating are obvious in many animals, and plants have chemical means of detecting pollen from the wrong species and rejecting it. Even if mating takes place, sperm from the wrong species may be unsuccessful in fertilizing the eggs of the female.

Some species are, however, sufficiently closely related that they will occasionally mate, especially if given no choice of a member of their own species (for example, dogs, coyotes, and jackals, mentioned in Chapter 5). In many such situations, however, the first-generation hybrids often fail to develop; experimental crosses between individuals belonging to different species often produce hybrids that die at an early stage of development, whereas most offspring of crosses between individuals of the same species develop to maturity. Sometimes hybrid individuals can survive, but at a much lower frequency than non-hybrids. Even when hybrids are viable, they are often sterile, and do not produce any offspring which could pass genes on to future generations; mules (which are hybrids produced by crosses between donkeys and horses) are a famous example. Complete inviability or sterility of the hybrids obviously isolates the two species.

The evolution of barriers to interbreeding

Although these different means of preventing interbreeding are familiar, it is a puzzle to understand how they could evolve. This is the key to understanding the origin of species. As Darwin pointed out in Chapter 9 of *The Origin of Species*, it is most unlikely that the inviability or infertility of interspecies hybrids could be the direct product of natural selection; there can be no advantage to an individual producing inviable or sterile offspring if hybridized with a different species. It would, of course, be advantageous to avoid mating with members of another species if the hybrid offspring are inviable or sterile, but it is difficult to see how there could be any such advantage in cases when the hybrids survive perfectly well. It therefore seems likely that most barriers to interbreeding between species are by-products of evolutionary changes that occurred after the populations became isolated from each other by being geographically or ecologically separated.

For example, imagine a species of Darwin's finch living on one of the Galapagos islands. Suppose that a small number of individuals manage to fly across to another island, previously unoccupied by this species, and successfully establish a new population. If such migration events are very rare, the new and the ancestral populations will evolve independently of each other. Under the processes of mutation, natural selection, and genetic drift, the genetic compositions of the two populations will diverge. These changes will be promoted by differences in the environments experienced by the populations, to which they become adapted. For example, the food plants available to a seed-eating species of bird differ from island to island, and even members of the same species of finch differ between islands in their beak sizes in ways which reflect differences in food abundance.

The tendency of populations of the same species to differ according to their geographical location, often in ways which are clearly adaptive, is called *geographical variation*. Obvious examples in the human species are the numerous minor physical differences between the races, as well as the smaller local differences in features such as skin pigmentation and stature. Such variability is found in many other species of animals and plants with wide geographical ranges. In a species that consists of a set of local populations, there is usually some migration of individuals between different locations. The amount of migration varies enormously between organisms; snails have very low migration rates, whereas organisms like birds or many flying insects are highly mobile. If migrant individuals can interbreed with members of the population in which they arrive, they will contribute their genetic makeup to this population. Migration is therefore a homogenizing force, opposing the tendency for local populations to diverge genetically by selection or genetic drift (see Chapter 2). Populations of a species will diverge more or less from each other, depending on the amount of migration, and on the evolutionary forces promoting differences between local populations. Strong selection can cause even adjacent populations to differ. For example, lead or copper mining produces soil contaminated with these metals, which are very toxic to most plants, but metal-tolerant forms have evolved on the polluted land surrounding many mines. In the absence of the metals, the tolerant plants grow poorly. Tolerant plants are therefore found only on or very close to the mines, and there is a sharp changeover to non-tolerant individuals at the boundaries.

In less extreme cases, gradual geographical changes in traits arise because migration blurs the differences caused by selection that varies geographically, in response to changes in environmental conditions. Many species of mammals living in the temperate zone of the northern hemisphere have larger body sizes in the north. Average body size changes more or less continuously from north to south, probably reflecting selection for a smaller ratio of surface area to volume in colder climates, where heat loss is a problem. Northern populations also tend to have shorter ears and limbs than southern populations, for similar reasons.

Differences between geographically separate populations of the same species do not necessarily require different types of selection. The same selection can sometimes lead to different responses. For example, as we described in Chapter 5, human populations in regions subject to malaria infections have different genetic mutations that confer resistance to malaria. There are multiple molecular paths to resistance. Different mutations that can cause resistance will occur by chance in different places, and it is largely luck which mutation comes to predominate in a given population. Differences between populations of the same species can also evolve even if there is no selection at all, as a result of the random process of genetic drift mentioned earlier. In many species, there are often marked genetic differences among different populations even for variants in DNA or protein sequences that have no effect on visible traits, and the human population is no exception to this. Even within Britain, there are differences in the frequencies of individuals with the A, B, and O blood groups, which are determined by variant forms of a single gene. The O blood group is more frequent in North Wales and Scotland than in the south of England, for example. Over wider areas, there are much greater differences in blood group frequencies. Blood group B has a frequency of over 30% in some parts of India, whereas it is virtually absent from native Americans.

There are many other such examples of geographical variation. Despite the visible differences between the major races, humans have no biological barriers to interbreeding between different populations or racial groups. In some species, however, populations from the extreme ends of a species range look different enough that they might well be regarded as different species, except for the fact that they are connected by a set of intergrading populations which interbreed with each other. There are even cases in which populations at opposite ends of a species range have diverged so much that they cannot interbreed; if the intermediates were to become extinct, they would constitute different species.

This illustrates an important point: on the theory of evolution, there must be intermediate stages in the development of reproductive isolation, and so we ought to observe at least some cases in which it is difficult to say whether or not a given pair of related populations belong to the same or different species. While this is inconvenient if we want to put things into cut-and-dried categories, it is a predictable outcome of evolution, and is evident in the natural world. There are many known examples of intermediate stages in the evolution of complete inability to interbreed between geographically separated populations. A particularly well-studied example is the American species of fruitfly, *Drosophila pseudoobscura*. This lives on the west coast of North and Central America, more or less continuously from Canada to Guatemala, but there is also an isolated population living near Bogotá, in Colombia. Flies from the Bogotá population look identical to those from other populations of the species, but their DNA sequences differ slightly. Since accumulation of sequence differences requires a long time, the Bogotá population was probably founded by a few migrant flies around 200,000 years ago. In the laboratory, flies from Bogotá will readily mate with *Drosophila pseudoobscura* flies from other populations; first-generation hybrid females are fully fertile, but hybrid males from the cross with non-Bogotá females as their mothers are sterile. No hybrid male sterility is observed in crosses among very different populations from the rest of the species range. If flies from the main population were introduced into Bogotá, they would presumably interbreed fairly freely with the Bogotá flies, and since the female hybrids are fertile, interbreeding could continue every generation. Thus the Bogotá population owes its distinctness purely to its geographic isolation. There is therefore no compelling reason to regard it as a separate species, although it is starting to develop reproductive isolation, as indicated by the sterility of the hybrid males.

It is relatively simple to understand why populations of the same species living in different places may come to diverge with respect to characteristics that adapt them to differences in the environment, as in the Galapagos finch example. It is less obvious why this leads to failure to interbreed. This may sometimes be a fairly direct by-product of adaptations to different environments. For example, two species of monkeyflower plants, *Mimulus lewisii* and *M. cardinalis*, grow in the mountains of the north-western USA. Like most monkeyflowers, *M. lewisii* is pollinated by bees, and its flowers show several adaptations for bee pollination (see the table on page 97). Unusually for a monkeyflower, *M. cardinalis* is pollinated by hummingbirds, and its flowers differ in several characteristics that promote pollination by hummingbirds. *M. cardinalis* thus probably evolved from a bee-pollinated ancestor, similar in appearance to *M. lewisii*, by a process of changing these flower characteristics.

Floral characteristics of two *Mimulus* species

Species	*M. lewisii*	*M. cardinalis*
Pollinators	bee	hummingbird
Flower size	small	large
Flower shape	wide, with 'landing platform'	narrow, tubular
Flower colour	pink	red
Nectar	moderate, high sugar	abundant, low sugar

The two monkeyflower species can be crossed experimentally, and the hybrids are healthy and fertile, yet in nature the species grow side by side without intermingling. Observations on pollinator behaviour in the wild show that, after visiting *M. lewisii*, bees rarely visit *M. cardinalis*, and a hummingbird that has visited *M. cardinalis* will rarely go on to a *M. lewisii* plant. To find out how pollinators would react to plants with intermediate flower traits, an artificially produced second-generation hybrid population, with a wide range of combinations of traits from the two parents, was planted in the wild. The trait that most strongly promoted isolation was flower colour, with red deterring bees and attracting visits by humming birds. Other traits affected one or the other of the two pollinators. A higher nectar volume per flower increased hummingbird visits, whereas flowers with larger petals were visited more often by bees. Intermediate forms between the two species had intermediate probabilities of being pollinated by bees versus hummingbirds, and hence intermediate degrees of isolation from the parent species. In this example, changes driven by natural selection as hummingbird pollination evolved have led to the *M. cardinalis* population becoming reproductively isolated from a closely related *M. lewisii* population.

Even though in most cases we do not know what force drove the divergence between closely related species and resulted in their reproductive isolation, the origin of reproductive isolation between a pair of geographically separated populations is not particularly surprising, if there have been independent evolutionary changes in two populations. Each alteration in the genetic composition of one population must either be favoured by selection in the population, or else must have such a slight effect on fitness that it can spread by genetic drift (discussed in Chapter 2, and at the end of this chapter). If a variant is spreading in a population because it has an advantage in adapting the population to its local environment, its spread will not be impeded by any harmful effects when combined (in hybrids) with genes from a different population which it never naturally encounters. There is no selection to maintain compatibility of mating behaviour between individuals from geographically or ecologically separated populations, or to maintain harmonious interactions that allow normal development, between genes that have come to differ in different populations. Like other characteristics that are not subject to selection to maintain them (such as the eyes of cave-dwelling animals), the ability to interbreed degenerates over time.

Given enough evolutionary divergence, complete reproductive isolation seems inevitable. It is no more surprising than the fact that electrical plugs of British design do not function in Continental European sockets, even though each type of plug functions perfectly with its own sockets. In human-designed machines where compatibility is desirable, constant efforts must be made to preserve it, for example in software for PC versus Macintosh computers. Genetic analyses of interspecies crosses show that different species really do contain different sets of genes which are dysfunctional when brought together in hybrids. As already mentioned, the first-generation male hybrids between many species of animals are sterile, while the females are fertile. Crosses are then possible between fertile hybrid females and either of the parental species. By testing the fertility of the male offspring of such crosses, we can study the genetic basis of the hybrid male sterility. This kind of study has been intensively carried out using *Drosophila* species; the results show clearly that the hybrid sterility is produced by interactions between different genes from the two species. In the case of the mainland versus Bogotá populations of *D. pseudoobscura*, for example, about 15 distinct genes which differ between the two populations seem to be involved in causing the sterility of hybrid males.

The time needed to produce sufficient differences between a pair of populations to make them incapable of interbreeding is very variable. In the *Drosophila pseudoobscura* example, 200,000 years (over a million generations) has produced only very incomplete isolation. In other cases, there is evidence for the very rapid evolution of barriers to interbreeding, as in the case of fish species of the cichlid family in Lake Victoria. Here, there are over 500 species apparently derived from one ancestral species, yet geological evidence shows that the lake has existed for only 14,600 years. Isolation between these species seems to be largely due to behavioural traits and coloration differences, and there is very little differentiation between the species in their DNA sequences. It seems to have taken around 1,000 years on average for a new species of this group to be produced, but other groups of fishes in the same lake have not evolved new species at such a high rate; typically, several tens of thousands of years seem to be needed for a new species to be formed.

Once two related populations have become completely isolated from each other by one or more barriers to interbreeding, their evolutionary fates will forever be independent of one other, and they will tend to diverge over time. One important cause of such divergence is natural selection; closely related species often differ in many structural and behavioural characteristics that adapt them to their different ways of life, as we have already described with the Galapagos finches. Sometimes, however, there are very few evident differences between related species. This is often the case with insects; for example, the *Drosophila* species *D. simulans* and

D. mauritiana both have very similar bodily structure, and differ externally only in the structure of the male genitalia. Nevertheless, they are true species, and are very reluctant to mate with each other. Similarly, it has recently been discovered that the common European pipistrelle bat is divided into two different species. They do not interbreed in nature, and differ in their calls as well as in their DNA sequences. Conversely, as we have already described, there are many examples of marked differences between populations of the same species, with no barriers to interbreeding.

These examples show that there is no absolute relationship between differences in easily observable characteristics and the strength of reproductive isolation between a pair of populations. Nor is the extent of differences between a pair of species very closely related to the time since they became reproductively isolated. This is illustrated by the striking differences among island species such as the Galapagos finches, which have evolved over a very short time-span compared with the amount of time that separates related South American species of birds, many of which differ far less (see Figure 13, Chapter 4). Similarly, in the fossil record, there are many examples of lineages showing little or no change over thousands or millions of years, followed by abrupt transitions to new forms, usually recognized as new species by palaeontologists.

Theoretical models, as well as laboratory experiments, show that intense selection can produce profound changes in a trait over 100 generations or less (see Chapter 5). For example, a population of the fruitfly *Drosophila melanogaster* has been selected artificially for an increase in the number of bristles on the flies' abdomens. Selection has produced a three-fold increase in average bristle number over 80 generations. This is about the same as the increase in the average size of the brain case between our earliest ape-like ancestors and ourselves, which took about 4 million years (roughly 200,000 generations). Conversely, traits will not change greatly, once a species living in a stable environment has had time to adapt to it. It is usually impossible to tell from the fossil record whether an observed 'sudden' evolutionary change implies the origin of a new species (which cannot interbreed with its progenitor), or simply involves a single lineage, evolving in response to environmental changes. In either case, there is no mystery in geologically rapid change.

Finally, what do species mean when there is asexual reproduction, which occurs in many single-celled organisms like bacteria? Here, the criterion of interbreeding is meaningless. For purposes of classification in these cases biologists simply use arbitrary measures of similarity, either based on characters of practical importance (such as the composition of bacterial cell walls), or increasingly on DNA sequence differences. Sufficiently similar individuals, which cluster together in regard to the characters that are used, are classed as the same species, whereas other groups of individuals that form a different cluster are assigned to different species.

Molecular evolution and divergence between species

Given the erratic relationship between the time since separation of a pair of species and their divergence in morphological characteristics, biologists are increasingly using information from the DNA sequences of different species to make inferences about their relationships.

Rather like comparisons of the spelling of the same word in different but related languages, we can see similarities as well as differences in the sequences of the same genes in different species. For example, *house* in English, *haus* in German, *huis* in Dutch, and *hus* in Danish all have the same meaning, and are pronounced very similarly. There are two types of differences between these words. First of all, there are changes of letter in a given position, as in the substitution of *a* for *o* in the second position between German and English. Second, there are additions and deletions of letters; the final *e* in English is missing from the other languages, and Danish lacks the *a* in the second place in German. Without more information on the historical relationships between the languages, it is difficult to be certain about the direction of these changes, although the fact that only English has a final *e* suggests quite strongly that this is a late addition, and the fact that *hus* is the shortest version suggests that a vowel has been lost from the Danish word. Given such comparisons of a large sample of words, the differences between different languages can be used to measure their relationships, and the differences correlate well with the time the languages have been diverging. American English is only a couple of hundred years separated from British English, but has diverged quite noticeably, including the development of different local versions. Dutch and German are more diverged, French and Italian even more so.

The same principle can be used for DNA sequences. In this case, changes due to insertions and deletions of individual letters in the DNA are rare in the portions of genes that code for proteins, since these will usually have major effects on the sequence of amino acids in the protein and render it non-functional. Between closely related species, most changes in the coding sequences of genes involve single changes to individual letters of the DNA sequence, such as changing a G to an A. An example is given in Figure 8, which shows sequences of portions of the melanocyte-stimulating hormone receptor gene from humans, chimpanzees, dogs, mice, and pigs.

By comparing the numbers of letters in the DNA by which the sequences of the same gene differ between a pair of different organisms, one can quantify their level of divergence precisely, which is difficult to do with morphological similarities and differences. Knowing the genetic code, we can see which of the differences alter the protein sequence corresponding to the gene in question (*replacement* changes), and which do not (*silent* changes). For instance, in the melanocyte-stimulating hormone receptor sequences, a simple count of the differences between the human and chimpanzee sequences in Figure 8 reveals four differences in the 120 DNA letters shown. For the entire sequences of the different species (omitting a small region with some additions and deletions of DNA letters), the numbers of differences from the human sequence are shown in the table below.

Human versus	Same amino acid (silent differences)	Different amino acid
Chimpanzee	17	9
Dog	134	53
Mouse	169	63
Pig	107	56

A recent study showed that the sequence divergence for 53 non-coding DNA sequences compared between humans and chimpanzees ranged between 0 and 2.6% of the total letters, and averaged only 1.24% (1.62% for the human and gorilla). These estimates show why it is now accepted that chimpanzees, rather than gorillas, are our closest living relatives. The differences are much greater if humans are compared with the orang-utan, and greater still if we are compared with baboons. More distantly related mammals, such as carnivores and rodents, differ at the sequence level much more than do different primates; mammals differ much more from birds than they do from each other, and so on. The patterns of relationships revealed by sequence comparisons are in broad agreement with what is expected from the times at which the major groups of animals and plants appear in the fossil record, as expected in the theory of evolution.

The table of sequence differences shows that silent changes are generally much more common than replacement changes, although even silent changes are rare between the most closely related species such as chimpanzees and humans. The obvious interpretation is that most changes to the amino acid sequence of a protein impair its function to some extent. As we described in Chapter 5, a small detrimental effect caused by a mutation will result in selection quickly eliminating the mutation from the population. Most mutations that change protein sequences therefore never contribute to evolutionary differences in gene sequences that accumulate between species. But there is also increasingly firm evidence that some amino acid sequence evolution is driven by selection acting on occasional favourable mutations, so that molecular adaptation occurs (see Chapter 5).

In contrast to the often detrimental effects of mutations changing amino acids, silent changes to the sequences of genes will have little or no effect on biological functions. It thus makes sense that most divergence in gene sequences between species are silent changes. But when a new silent mutation appears in a population, it is just a single copy among thousands or millions of copies of the gene in question (two in each individual in the population). How does such a mutation spread through the population if it does not confer any selective advantage to its carrier? The answer is that random changes in the frequencies of alternative variants (genetic drift) take place in finite populations, a concept we introduced briefly in Chapter 2.

This process works as follows. Suppose that we study a population of the fruitfly *Drosophila melanogaster*. For the population to be maintained, each adult must contribute on average two descendants to the next generation. Suppose that the population varies in eye colour, with some individuals carrying a mutant gene that makes the eyes bright red while the non-mutant version of this gene makes all the other flies' eyes the normal dull red. If individuals with either type of gene have the same average number of offspring, there is no selection on eye colour; it is said to be *neutral* in its effects. Because of this neutrality with respect to selection, the genes of the next generation will be drawn randomly from the parental population (Figure 18). Some individuals may have no offspring, while others may happen by chance to have more than the average of two offspring. This means that the frequency of the mutant gene in the progeny generation will not be the same as its frequency among the parents, since it is extremely unlikely that individuals with and without the mutant gene contribute exactly the same numbers of offspring. Over the generations, there will thus be continual random fluctuations in the composition of the population, until sooner or later either all members of the population have the gene for bright red eyes, or else it is lost from the population and they all have the alternative version of the gene. In a small population, genetic drift is fast, and it will not take long until all members of the population become the same. This will take much longer in a large population.

This illustrates two effects of genetic drift. First, while a new variant is drifting to eventual loss or to a frequency of 100% (*fixation*), the character affected by the gene is variable within the population. The input of new neutral variants by mutation and the changes in variant frequencies (and, from time to time, loss of variant genes) by drift determines the variability in the population. Examination of DNA sequences of the same gene from different individuals from a population reveals variability at silent sites due to this process, as we mentioned in Chapter 5.

A second effect of genetic drift is that a selectively neutral variant that is initially very rare has some chance of spreading throughout the whole population and replacing alternative variants, although it has a much greater chance of being lost. Genetic drift thus leads to evolutionary divergence between isolated populations, even without any selection promoting the changes. This is a very slow process. Its rate depends on the rate at which new neutral mutations arise, as well as the rate at which genetic drift leads to replacement of one version of a gene by a new one. Remarkably, it turns out that the rate of DNA sequence divergence between a pair of species depends only on the rate of mutation per DNA letter (the frequency with which a particular letter in a parent is mutant in the copy that is passed to an offspring). An intuitive explanation for this is that, if

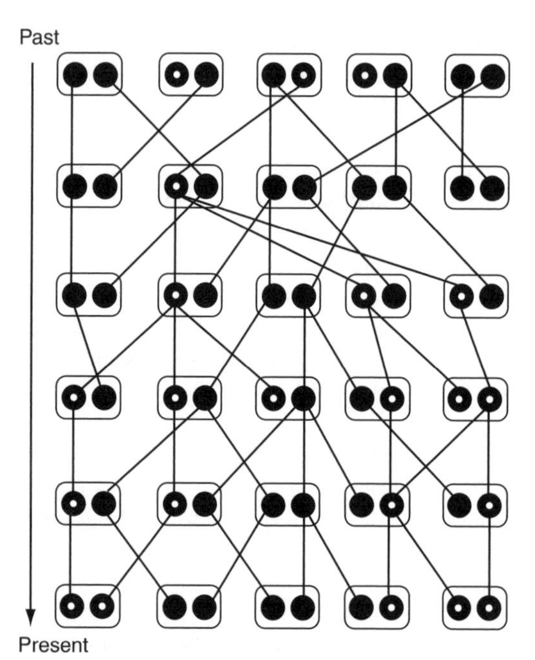

18. Genetic drift. The process of genetic drift of a single gene over six generations, in a population of five individuals. Each individual (symbolized by an open shape) has two copies of the gene, one from each parent. The different DNA sequences of the individuals' gene copies are not shown in detail, but are symbolized by black discs with or without a white spot. The white spots might correspond to the variant gene causing bright red eye colour, and the black discs to the variant with dull red eye colour, in the *Drosophila* example given in the text. In the first generation, three individuals have one of the white spot type of the gene and one of the black type. Thus 30% of the genes in the population have the white spot. The figure shows the lines of descent of the genes in each generation (we assume for convenience that individuals can reproduce as either male or female, as is true for many hermaphroditic species of plants, such as tomatoes, and some animals, such as earthworms). Some individuals happen by chance to have more offspring than others, while other have less, or may even leave no surviving descendants (e.g. the individual shown at the right in generation 2). The numbers of white spot and black gene copies therefore fluctuate from each generation to the next. In the third generation, three individuals all inherit the white spot gene copy from the single individual carrying one such gene in generation 2, so this type of gene goes from 10% to 30%; in the next generation it is 50%, and so on.

no selection is acting, nothing affects the number of mutational differences between a pair of species except the rate at which mutations appear in the sequence and the amount of time since the species' last common ancestor. A large population has more new mutations per generation, simply because there are more individuals in which a mutation might happen. But genetic drift happens faster in a small population, as explained above. It turns out that the two opposing effects of population size cancel out exactly, and so the mutation rate determines the rate of divergence.

This theoretical result has important implications for our ability to determine the relationships between different species. It implies that neutral changes accumulate in a gene as time goes on, at a rate that depends on the gene's mutation rate (the molecular clock principle, which we mentioned, but did not explain, in Chapter 3). Sequence changes in genes are therefore likely to take place in a much more clock-like fashion than changes in traits subject to selection. Rates of morphological changes depend strongly on environmental changes, and variable rates and reversals of direction can occur.

Even the molecular clock is not very precise. Rates of molecular evolution can change over time within the same lineage, as well as between different lineages. Nevertheless, use of the molecular clock allows biologists to roughly date the divergence between species for which there is no fossil evidence. To calibrate the clock, one needs sequences from the closest available species whose divergence dates are known. One of the most important applications of this method has been to date the timing of the split between the lineage giving rise to modern humans and the one leading to chimpanzees and gorillas, for which no independent fossil evidence is available. Use of the molecular clock with a large number of gene sequences has enabled a date of 6 or 7 million years to be estimated with considerable confidence. Because the rate of neutral sequence evolution depends on the mutation rate, the clock is exceedingly slow, since the rate at which single letters in the DNA change by

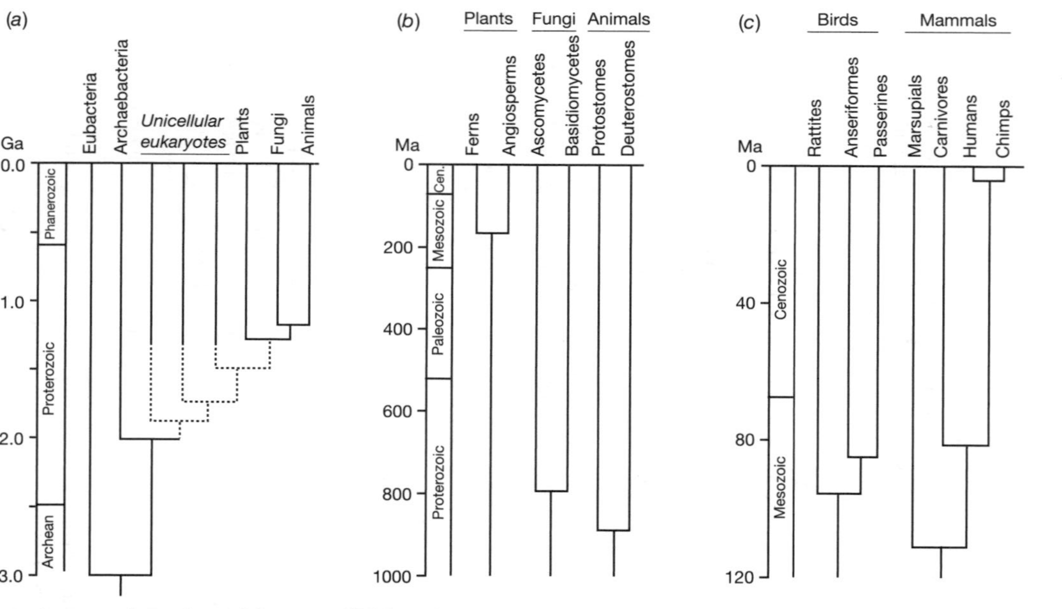

19. A recent chronology of the tree of life based on DNA sequence differences, with the estimated divergence dates between groups. Part (a) shows all organisms (eubacteria and archaebacteria are the two great divisions of bacteria), (b) shows multicellular organisms (angiosperms are flowering plants; ascomycetes and basidiomycetes are the two major types of fungi), and (c) shows bird and mammal groups (rattites are ostriches and their relatives; anseriformes are ducks and their relatives; and passerines are the song-birds).

mutation is very low. The fact that approximately 1% of the DNA letters differ between humans and chimpanzees corresponds to a single letter changing only once in over a billion years. This is consistent with experimental measurements of mutation rates.

A molecular clock is also found to apply to the amino acid sequence of proteins. As already mentioned, protein sequences evolve more slowly than silent DNA differences, and are therefore useful for the difficult task of comparing species that diverged a very long time ago. Between such species, multiple changes will have occurred at some sites in their DNA sequences, so that it becomes impossible to count accurately the number of mutations that have happened. Scientists who are interested in reconstructing the times of divergence between the major groups of living forms therefore use data from slowly evolving molecules (Figure 19). Such dates are, of course, rough estimates, but the accumulation of estimates from many different genes can improve the accuracy of the procedure. Judicious use of sequence information from genes that evolve at different rates is allowing evolutionary biologists to form a picture of the relationships between groups of organisms whose last common ancestors lived a billion or more years ago. In other words, we are getting close to reconstructing the genealogical tree of life.

Chapter 7
Some difficult problems

As the theory of evolution has become increasingly well understood and tested by biologists, new questions have arisen. Not all problems have been solved, and there is still debate about old questions as well as new ones. In this chapter, we describe some examples of biological phenomena that are apparently difficult to explain. Some of these were dealt with by Darwin himself, others have been the subject of later research.

How can complex adaptations evolve?

Critics of the theory of evolution by natural selection frequently raise the difficulty of evolving complex biological structures, from protein molecules through single cells to eyes and brains. How can a fully functioning and beautifully adapted piece of biological machinery be produced purely by selection acting on chance mutations? The key to understanding how this can happen is expressed in another meaning of the word 'adapt'. In the evolution of organisms and their complex machinery, many aspects are modified (adapted) versions of pre-existing structures, just as when machines are made by engineers. In making complex machines and devices, less elegant initial models are refined over the course of time and diversified (adapted) to new, sometimes unanticipated, uses. The evolution of the total knee replacement is a good example of the process by which a crude initial solution to a problem was good enough to be useful, but was successively adapted to work better and better. Just as in biological evolution, many early designs were developed that seem poor by today's standards, yet each was an improvement on the ones before, and could be used by knee surgeons. These each played their roles as stages in the evolution of modern, complex artificial knees.

This process of successive adaptation of 'designs' is like climbing a hill in a thick fog. Even without a goal of reaching the top (or even without knowing where it is), if one follows a simple rule – each step goes uphill – one will move closer and closer to the summit (or at least to a local top). Simply by making a structure work better in one way or another, the end result is an improved design, without a Designer being necessary. In engineering, improved design is often the result of many contributions from different engineers over the evolution of a machine, and early car designers would have been astonished at modern cars. In natural evolution, it results from what has been called 'tinkering' with the organism, with minor changes that make their possessors survive or reproduce better than others. In the evolution of a complex structure, several different traits must, of course, evolve simultaneously, so that the different parts of the structure are well adapted to function as a whole. We saw in Chapter 5 that advantageous traits can spread through a population over a short time, relative to the time available for major evolutionary changes, even if they are initially very rare. A succession of small changes to a structure that already works, but can be improved, can therefore produce large evolutionary changes. After many thousands of years, the radical transformation of even a complex structure is not difficult to imagine. After enough time, the structure will differ from its ancestral state in many different ways, so that individuals in the descendant population would have combinations of characteristics never seen in the ancestral population, just as modern cars have many differences from early cars. This is not just a theoretical possibility: as we described in Chapter 5, animal and plant breeders routinely accomplish this by artificial selection. There is thus no difficulty in seeing how natural selection can cause the evolution of highly complex characters, made up of numerous mutually adjusted components.

The evolution of protein molecules is sometimes posed as an especially difficult problem. Proteins are complex structures whose parts must interact to function properly (many proteins must also interact with other proteins and other molecules, including DNA in some cases). The theory of evolution must certainly be capable of accounting for protein evolution. There are 20 different kinds of amino acids, so the chance that the right one would appear at a particular site in a protein molecule 100 amino acids long (shorter than many real proteins) is 1 in 20. The chance is evidently vanishingly small that, if 100 amino acids were randomly thrown together, each position in the sequence would have the right amino acid, and a working protein would form. It has therefore been claimed that the chance of assembling a functioning protein is similar to that of an airliner being assembled by a tornado blowing through a scrapyard. It is true that a functioning protein could not be assembled by randomly picking an amino acid for each position in the sequence. But, as the explanation given above makes clear, natural selection does not work like this. Proteins probably started as short chains of a few amino acids that could cause reactions to go a bit faster, and were successively improved as they evolved. There is no need to worry about the many millions of potential non-functional sequences that will never exist, provided that protein sequences during evolution started off catalysing reactions better than when no protein is present, and then got successively better over evolutionary time. It is easy to see in principle how successive stepwise changes, each one changing the sequence or adding to its length, could improve a protein.

Our knowledge about how proteins function supports this. The part of a protein that is essential for its chemical activity is often only a very small part of its sequence. A typical enzyme has just a handful of amino acids that physically interact with the chemical that is to be changed by the enzyme. Most of the rest of the protein chain simply provides a scaffold that supports the structure of the part involved in this interaction. This implies that the functioning of a protein depends critically on only a relatively small set of amino acids, so that a new function could evolve by a small number of changes to the sequence of the protein. This has been verified by numerous experiments in which artificially induced changes to protein sequences have been subjected to selection for new activities. It has proved surprisingly easy to produce quite radical shifts in the biological activity of proteins by these means, sometimes just by a change in a single amino acid, and there are similar examples among naturally evolved changes.

A similar answer can be given to the question of how it is possible for pathways of successive enzyme reactions to evolve, such as those which make chemicals that organisms need (see Chapter 3). One might think that, even if the end-products are useful, it would be impossible to evolve such pathways, since evolution has no foresight and cannot build up a chain of enzyme reactions until its function is complete. Again, the solution to this apparent riddle is simple. Many useful chemicals were probably present in the environment of early organisms. As life evolved, these would become scarce. An organism that could change a similar chemical into the useful one would benefit, and so an enzyme could evolve to catalyse that change. The useful chemical would now be synthesized from the related one. Thus a short biosynthetic pathway, with a precursor and a product, would be favoured. By successive steps like this, pathways could evolve – backwards from their end-products – to build up the chemicals organisms need.

If complex adaptations really evolve in steps, as evolutionary biologists propose, we should be able find evidence for intermediate stages in the evolution of such characters. There are two sources of such evidence: the existence of intermediates in the fossil record, and present-day species that show intermediate stages between simple and more advanced states. In Chapter 4, we described examples of intermediate fossils linking very different forms; these support the principle of stepwise evolutionary changes. Of course, in many cases there is a complete absence of intermediates, especially as we go further back in time. In particular, the major divisions of multicellular animals, including molluscs, arthropods, and vertebrates, nearly all appeared rather suddenly in the Cambrian (more than 500 million years ago), with virtually no fossil evidence concerning their ancestors. Recent DNA sequence studies of the relationships between them suggests strongly that these groups were already separate lineages long before the Cambrian era (Figure 19), but we simply have no information on what they looked like, probably because they were soft-bodied and hence unlikely to fossilize. But the incompleteness of the fossil record does not mean that intermediates did not exist. New intermediates are constantly being discovered. A recent one is a 125-million-year-old mammalian fossil from China with features similar to those of modern placental mammals, but more than 40 million years older than the oldest previously known fossil of this kind.

The other type of evidence, from comparisons of living forms, is our only source of information on features that do not fossilize. A simple but compelling example is provided by flight, as pointed out by Darwin in Chapter 6 of *The Origin of Species*. There are no fossils connecting bats with other mammals; the first bat fossils, found in deposits over 60 million years old, have the same highly modified limbs as modern bats. But there are several examples of modern mammals which have the ability to glide but cannot fly. The most familiar are the 'flying' squirrels, which are very similar to ordinary squirrels except for flaps of skin connecting their fore- and hindlimbs. These act as a crude wing, which allows the squirrels to glide some distance if they launch themselves into the air. Similar adaptations to gliding have evolved independently in other mammals, including the so-called flying lemurs (which are not true lemurs, and are not related to flying squirrels), and in the marsupial sugar-gliders. Gliding species of lizards, snakes, and frogs are also known. It is easy to imagine how being able to glide reduces the risk of a tree-living animal being caught and eaten by a predator, and that gliding could evolve by a gradual modification of the body of an animal that jumps from branch to branch. A gradual increase in the area of skin used for gliding, and modifications of the forelimbs to support such an increase, would clearly be advantageous. The flying lemur has a large extensible membrane stretching from head to tail. This is very close to the wings of bats, although the animals can only glide, not fly. Once a wing structure that allows highly efficient gliding has evolved, the development of wing musculature to produce power strokes can readily be envisaged.

The evolution of eyes is another example, also considered by Darwin. The vertebrate eye is a highly complex structure, with its light-sensitive cells in the retina, the transparent cornea and lens that allow the image to be focused on the retina, and muscles that adjust the focus. All vertebrate animals have essentially the same design of eye, but with many variations of detail adapted to different modes of life. How could such a complex piece of machinery evolve, when a lens is apparently useless without without a retina, and vice versa? The answer is that a retina is certainly not useless without a lens. Many types of invertebrate animals have simple eyes, with no lens. Such animals do not need to see clearly. It is enough to perceive light and dark in order to detect predators. In fact, a whole series of intermediates between simple light-sensitive receptors and various types of complex devices that produce images of the world can be seen in different groups of animals (Figure 20). Even single-celled eukaryotes are capable of detecting and responding to light, by means of receptors composed of a cluster of molecules of the light-sensitive protein rhodopsin. Rhodopsin is used in all animal eyes, and is also found in bacteria. Starting with this simple ability of cells to detect light, it is easy to imagine a series of steps in which increased light-capturing abilities evolve step by step, leading

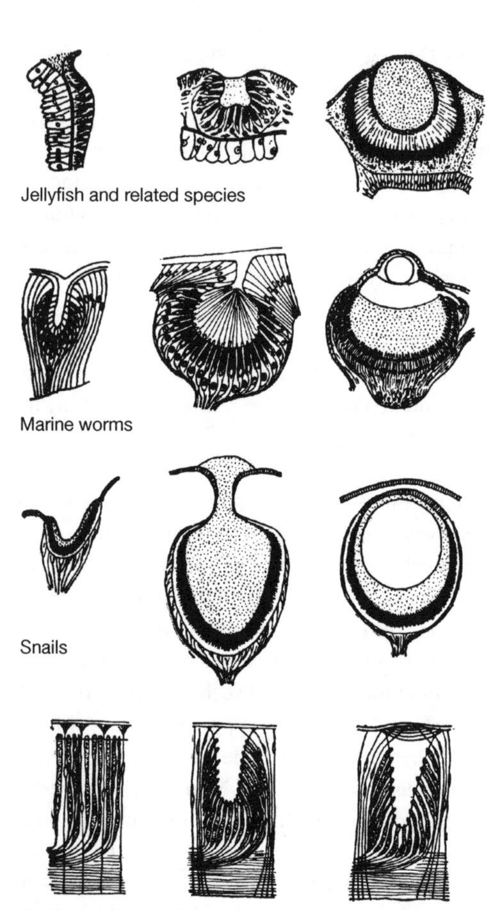

20. Eyes of a variety of invertebrate animals. From left to right, each row shows successively more advanced types of eyes, possessed by different species within a given group. For example, in the marine worms (second row), the left-hand eye consists simply of a few light-sensitive and pigment cells with a transparent cone projecting into their midst. The middle eye has a chamber filled with transparent jelly and a retina with a large number of light-sensitive cells. The right-hand eye has a spherical lens in front of the chamber, and many more light receptors.

eventually to a focusable lens that produces a sharp image. As Darwin put it:

> In living bodies, variation will cause the slight alterations, . . . and natural selection will pick out with unerring skill each improvement. Let this process go on for millions of years; and during each year on millions of individuals of many kinds; and may we not believe that a living optical instrument might thus be formed . . . superior to one of glass?

Why do we age?

The bodies of young adults as a whole strike us, like the eye, as near-perfect pieces of biological machinery. The opposite problem to explaining this near-perfection is presented by the fact that it is not maintained for very long during life. Why does evolution allow this to happen? The decline of a near-perfect creature into a feeble shadow of itself as a result of ageing has been a favourite topic of the poets, especially when they foresee it happening to their lovers:

> Then of thy beauty do I question make,
> That thou among the wastes of time must go,
> Since sweets and beauties do themselves forsake,
> And die as fast as they see others grow;
>   And nothing 'gainst Time's scythe can make defence
>   Save breed to brave him when he takes thee hence.
>
> From 'Sonnet 12' by William Shakespeare

Ageing is, of course, not confined to humans; it has been observed in virtually every plant or animal. To measure ageing, we can study many individuals kept in a protected environment, where 'external' causes of mortality such as predation have been removed, so that individuals live for much longer than in nature. Following them over time, we can determine the probabilities of death at different ages. Mortality is usually high for very young individuals, even in protected conditions; it declines as juveniles become older and larger, but then increases again after adulthood. In most species that have been studied carefully, the adult death rate increases steadily with age. Mortality patterns, however, differ greatly in different species. Small, short-lived organisms such as mice have much higher mortality rates at relatively young ages than large, long-lived organisms like ourselves.

This senescent increase in mortality reflects the deterioration of multiple biological functions with advancing age: almost everything seems to get worse, from muscular strength to mental ability. The nearly universal occurrence of ageing in multicellular organisms (which seems like a kind of degeneration) may seem to be a severe difficulty for evolutionary theory – contradicting the idea that natural selection causes the evolution of adaptation. One answer to this is that adaptation is never perfect. Ageing is partly an unavoidable consequence of cumulative damage to the systems necessary for continued survival, and selection probably simply cannot prevent this. Indeed, the annual chance of failure of complicated machines, such as cars, also increases with age, quite similarly to the mortality of living organisms.

But this cannot be the whole story. Single-celled organisms like bacteria reproduce simply by dividing into daughter cells, and the lineages of cells produced by these divisions have persisted over billions of years. They do not senesce, but continually break down damaged components and replace them with new ones. They can continue to propagate indefinitely, provided that harmful mutations are removed by selection. This is also possible for artificially cultured cells of some organisms, such as fruitflies. The reproductive cell lineages of multicellular organisms are also perpetuated every generation, so why could repair processes not be maintained for the whole organism? Why do most of our body systems show some senescent decline? For example, mammals' teeth wear out with advancing age, eventually leading to death from starvation in nature. This is not inevitable; reptiles' teeth are renewed from time to time. The different rates of ageing of different species reflects different effectiveness of repair processes and the extent to which these are maintained with advancing age: a mouse can expect to live for about three years at best, whereas a human can live for more than 80 years. These species differences indicates that ageing evolves. Ageing therefore demands an evolutionary explanation.

We saw in Chapter 5 that natural selection on multicellular organisms works in terms of differences in individuals' contributions to the next generation, through differences in the numbers of offspring they produce, as well as in their chances of survival. Furthermore, all individuals have some risk of death from accidents, diseases, and predation. Even if the chance of death from these causes is age-independent, the chance of surviving goes down as age increases, in ourselves, as for cars: if the probability of survival from one year to the next is 90%, the chance of survival over five years is 60%, but over 50 years it is only 0.5%. Selection therefore favours survival and reproduction early rather than late in life, simply because, on average, more individuals will be alive to experience the good effects. The greater the mortality due to accidents, diseases, and predation, the more strongly will selection favour improvements early in life, relative to later on, since few individuals can survive to late ages if the death rate from these external causes is high.

This argument suggests that ageing evolves because of the greater selective value of variants with favourable effects on survival or fertility early in life, compared with variants that act late. The concept is similar to the familiar idea of life insurance: it costs less to buy a given amount of insurance if you are young, because you most likely have many years of paying ahead. There are two main ways in which natural selection might work to cause ageing. The argument used above shows that mutations with harmful effects will be most strongly selected against if they express their effects early in life. The first way that selection can cause ageing is to keep early-acting mutations rare in populations, while allowing ones with effects late in life to become common. Indeed, many common human genetic diseases are due to mutations whose harmful effects appear late in life, such as those involved in Alzheimer's disease. Second, variants that have beneficial effects early in life will be more likely to spread through the population than those whose good effects come only in old age. Improvements to the early stages of life can evolve, even if these benefits come at the expense of harmful side-effects later on. For example, higher levels of some reproductive hormones may enhance women's fertility early in life, but at the risk of later breast and ovarian cancer. Experiments confirm these predictions. For example, one can keep populations of the fruitfly *Drosophila melanogaster* by breeding only from very old individuals. In a few generations, these populations evolve slower ageing, but at the expense of reduced reproductive success early in life.

The evolutionary theory of ageing predicts that species with low externally caused death rates should evolve low rates of ageing and longer life-spans, compared with species with higher external death rates. There is indeed a strong relation between body size and the rate of ageing, smaller species of animals tending to age much faster than larger ones, and to reproduce earlier. This probably reflects the greater vulnerability of many small animals to accidents and predation. Between species with similar body sizes, striking differences in rates of ageing between animals with different mortality rates in the wild often make sense when we consider their risks of predation. Many flying creatures are notable for longevity, which makes sense because flying is a good defence against many predators. A fairly small creature like a parrot can have a life-span longer than that of a human being. Bats live much longer than terrestrial mammals like mice with comparable body weights.

We ourselves may be an example of evolution of a slower rate of ageing. Our closest relatives, the chimpanzees, rarely live beyond 50 years even in captivity, and start reproducing earlier in life than humans, at an average of 11 years of age. Humans have therefore probably substantially reduced their rate of ageing since diverging from our common ancestor with apes, and postponed reproductive maturity. These changes are probably due to increased intelligence and ability to cooperate, which reduced vulnerability to external causes of death and reduced the advantage of reproducing early. A change in the relative advantages of early versus late reproduction can be detected and even measured in present-day societies. Industrialization has led to a dramatic decline in mortality rates among adults, which is evident in census data. This changes the natural selection affecting the ageing process in human populations. Consider the degenerative brain disorder Huntington's disease, which is caused by a rare mutant gene. This disease has a late age of onset (in the 30s or later). In a population with high mortality due to disease and malnutrition, few individuals survive to their 40s, and carriers of Huntington's disease have on average only slightly (9%) fewer offspring than unaffected individuals. In industrialized societies, with low mortality, people quite often have children at ages when the disease could appear, and in consequence affected people have on average 15% fewer children than unaffected individuals. If current conditions continue, selection will gradually reduce the frequencies of mutant genes with effects expressed late in reproductive life, and the survival rates of older individuals will increase. Rare genes with major effects like Huntington's disease have only a minor effect on the population as a whole, but many other diseases which are under at least partially genetic control predominantly afflict middle-aged and elderly people, including heart disease and cancer. We may expect the incidence of these genes to decline over time due to this natural selection. If the low death rates characteristic of industrialized societies persist for several centuries (a big if), there will be a slow but steady genetic change towards lower rates of ageing.

The evolution of sterile social castes

Another problem for evolutionary theory is posed by the existence of sterile individuals in a number of types of social animals. In social

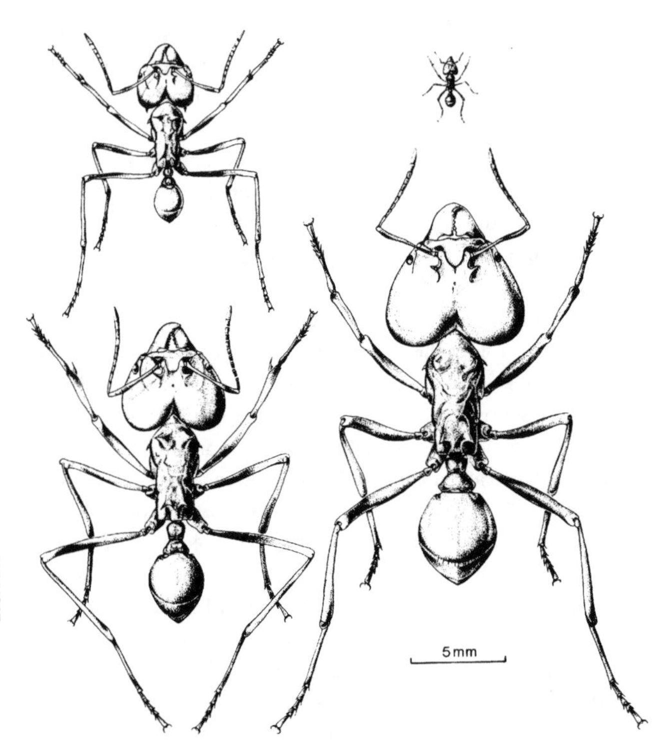

21. Castes of workers of the leaf-cutting ant *Atta*, all from the same colony. The tiny worker at the top right tends the fungus gardens cultivated by this species. The giant individuals are soldiers, who guard the nest.

wasps, bees, and ants, some of the females in a nest are workers, who do not reproduce. Reproductive females are a small minority within the colony (often just a single queen); the worker females look after the queens' offspring and maintain and provide for the nest. In the other main group of social insects, the termites, both males and females can behave as workers. In the advanced social insects, there are often several different 'castes', which perform very different roles and are distinguished by differences in behaviour, size, and body structure (Figure 21).

A remarkable recent discovery is that a few species of communally nesting mammals have social organizations resembling these insects, with the majority of inhabitants of a nest being sterile. The most famous is the naked mole rat, a species of burrowing rodent inhabiting desert areas of southern Africa. There may be several dozen inhabitants of a nest, with only a single reproductive female. If she dies, there is a struggle to replace her among some of the other females, in which one emerges victorious. Systems of social animals with sterile workers have thus evolved in quite different groups of animals. These species pose apparent problems for the theory of natural selection. How can individuals evolve to forego reproduction? How can the often very extreme adaptations of the castes of workers to their specialized roles have evolved, since the workers themselves do not reproduce, and so cannot be subject directly to natural selection?

These questions were raised, and partially answered, by Darwin in *The Origin of Species*. The answers lie in the fact that the members of a social group, such as a naked mole rat nest, or ant nest, are generally close relatives, often sharing the same mother and father. A genetic variant that causes its carriers to forgo their own reproductive success to help raise its relatives may help the relatives' genes pass to the next generation, and the relatives' genes are often (because of relatedness) the same as the helper individual's own genes (in the case of a brother and sister, if one individual has a genetic variant inherited from one parent, the chance is one-half that the variant will be present in the other). If the sacrifice by sterile individuals results in a sufficient increase in the numbers of surviving and reproductively successful relatives, the increase in number of copies of the 'worker gene' can outweigh the decrease due to their own lost reproductive success. The increase needed to outweigh the loss is smaller the closer the degree of relationship. J. B. S. Haldane once stated that 'I would lay down my life for two brothers or eight cousins'.

This principle of *kin selection* provides a framework for

understanding the origins of sterility in social animals, and modern research has shown that it can account for many details of animal societies, including those with less extreme features than sterile castes. For example, in some bird species, juvenile males do not attempt to mate, but remain as 'helpers' at their parental nest while younger siblings are being cared for. Similarly, wild dogs baby-sit a pack's young while other pack members go out hunting.

The question of how the differences between castes of sterile workers arise is slightly different, but has a related answer. Development as a member of a particular caste of worker is controlled by environmental cues, such as the amount and quality of food provided to the individual while a larva. However, the ability to respond to such cues is genetically determined. A certain genetic variant might confer the potentiality of a sterile member of an ant colony to develop as, say, a soldier (with bigger jaws than ordinary workers) rather than a worker. If a colony with soldiers is better defended against enemies, and if colonies with the variant can produce more reproductives on average, the variant will increase the success of its colony. If the reproductively active members of the colony are close relatives of the workers, the genetic variant that induces some workers to become soldiers will be transmitted by the colony via queens and males founding new colonies. Selection can thus act to increase the representation of this variant among colonies in the species.

These ideas also illuminate the evolution of multicellular organisms from single-celled ancestors. The cells produced from the fusion of an egg and sperm remain associated, and most of them lose the ability to become sex cells and contribute directly to the next generation. Since the cells involved are all genetically identical, this would be advantageous if survival and reproduction were sufficiently increased in the group of associated cells, compared to the single-celled alternative. The non-reproducing cells 'sacrifice' their own reproduction for the benefit of the community of cells. Some are doomed to die during the developmental process, as tissues form and dissolve, and many of them lose the potential to divide, as we explained when discussing the evolution of ageing. The serious consequences for organisms when cells regain the ability to divide without regard to the organism are manifested in cancer. The differentiation of cells into different types during development is analogous to the differentiation of castes in social insects.

The origin of living cells and the origin of human consciousness

Two other major and largely unsolved problems in evolution, at the opposite extremes of the history of life, are the origin of the basic features of living cells and the origin of human consciousness. In contrast to the questions we have just been discussing, these are unique events in the history of life. Their uniqueness means that we cannot use comparisons among living species to make firm inferences about how they might have occurred. In addition, the lack of any fossil record for the very early history of life or for human behaviour means that we have no direct information about the sequences of events involved. This does not, of course, prevent us from making guesses about what these might have been, but such guesses cannot be tested in the ways we have described for ideas about other evolutionary problems.

In the case of the origin of life, the aim of much current research is to find conditions resembling those which prevailed early in the Earth's history, which allow the purely chemical assembly of molecules that can then replicate themselves, just as the DNA of our own cells is copied during cell division. Once such self-replicating molecules have been formed, it is easy to imagine how competition between different types of molecule could result in the evolution of more accurate and faster replicating molecules, that is natural selection would act to improve them. Chemists have been very successful in showing that the basic chemical building blocks of life (sugars, fats, amino acids, and the constituents of DNA and RNA) can be formed by subjecting solutions of simpler molecules (of the type that are likely to have been present in the oceans of the early Earth) to electric sparks and ultra-violet irradiation. There has been limited progress in showing how these can be assembled into still more complex molecules that resemble RNA or DNA, and even more limited success in getting such molecules to self-replicate, so we are still far from achieving the desired goals (but progress is constantly being made). Furthermore, once this goal is achieved, the question of how to evolve a primitive genetic code that allows a short RNA or DNA sequence to determine the sequence of a simple protein chain must be solved. There are many ideas, but as yet no definitive solutions to this problem.

Similarly, we can only make guesses about the evolution of human consciousness. It is even difficult to state the nature of the problem clearly, since consciousness is notoriously hard to define precisely. Most people would not regard a newborn baby as conscious; few would dispute that a two-year-old child is conscious. The extent to which animals are conscious is fiercely debated, but pet-lovers are well aware of the ability of dogs and cats to react to the wishes and moods of their owners. Pets even seem to be able to manipulate their owners into doing what they want. Consciousness is thus probably a matter of degree, not kind, so that in principle there is little difficulty in imagining a gradual intensification of self-awareness and ability to communicate during the evolution of our ancestors. Some would regard language ability as the strongest criterion for possession of true consciousness; even this develops gradually with age in infants, albeit with astonishing speed. Furthermore, there are clear indications of rudimentary language abilities in animals such as parrots and chimpanzees, who can be taught to communicate simple pieces of information. The gap between ourselves and higher animals is more apparent than real.

Although we know nothing of the details of the selective forces driving the evolution of human mental and language abilities, which evidently far exceed those of any other animals, there is nothing particularly mysterious in explaining them in evolutionary terms. Biologists are making rapid progress in understanding the functioning of the brain, and there is little doubt that all forms of mental activity are explicable in terms of the activities of nerve cells in the brain. These activities must be subject to control by genes that specify the development and functioning of the brain; like any other genes, these will be liable to mutation, leading to variation on which selection can act. This is no longer pure hypothesis: mutations have been found which lead to deficiencies in specific aspects of grammar in the speech of their carriers, leading to identification of a gene involved in the control of some aspects of grammar. Even the mutation in its DNA sequence that causes the difference from normal is known.

Chapter 8
Afterword

What have we learned about evolution in the 140 years since Darwin and Wallace first published their ideas? As we have seen, the modern view is remarkably close in many ways to theirs, with a strong consensus that natural selection is the major force guiding the evolution of structures, functions, and behaviours. The chief difference is that two advances mean that the process of evolution through selection acting on random mutations of the genetic material is now much more credible than it was at the beginning of the 20th century. First, we have a much richer body of data demonstrating the action of natural selection at every level of biological organization, from protein molecules to complex behaviour patterns. Second, we also now understand the mechanism of inheritance, which was a mystery to Darwin and Wallace. Many important aspects of heredity are now understood in detail, from how the genetic information is stored in the DNA, to how it controls the characteristics of the organism through the intermediacy of the proteins that it specifies and by regulating their levels of production. In addition, we now understand that many changes in DNA sequences have little or no effect on the functioning of the organism, so that evolutionary changes in sequences can occur by the random process of genetic drift. The technology of DNA sequencing enables us to study variation and evolution of the genetic material itself, and to use sequence differences to reconstruct the genealogical relationships between species.

This knowledge of heredity, and our understanding that natural selection drives the evolution of organisms' physical and behavioural characteristics, does not imply rigid genetic determination of all aspects of such characteristics. The genes lay down only the potential range of traits that an organism can exhibit; the traits which are actually expressed often depend on the particular environment in which an organism finds itself. In higher animals, learning plays a major role in behaviour, but the range of behaviour that can be learned is limited by the animal's brain structure, which is in turn limited by the animal's genetic make-up. This certainly applies across species: no dog will ever learn to talk (nor will humans be able to smell rabbits at a distance). Among humans, there is strong evidence for the involvement of both genetic and environmental factors in causing differences in mental characteristics; it would be astonishing if this were not the case in our species, as in other animals. Most variability among humans is between individuals within local populations, and differences between populations are far fewer. There is thus no basis for treating racial groups as homogeneous, distinct entities, much less for ascribing genetic 'superiority' to any one of them. This is an example of how science can provide knowledge to inform people's decisions on social and moral issues, although it cannot prescribe those decisions.

The characteristics which we regard as most human, such as our ability to talk and to think symbolically, as well as the feelings that guide our family and social relationships, must reflect a long process of natural selection that started tens of millions of years ago, when our ancestors started living in social groups. As we saw in Chapter 7, animals that live in social groups can evolve behaviour patterns that are not purely selfish, in the sense of promoting an individual's survival or reproductive success at the expense of another's. It is tempting to think that such characteristics as a sense of fairness towards others form part of our evolutionary heritage as a social animal, just as parental care of children surely represents evolved behaviour similar to that exhibited by many other animals.

We emphasize again that this does not mean that all details of people's behaviour are genetically controlled, or that they represent characteristics that increase human fitness. Moreover, there is great difficulty in conducting rigorous tests of evolutionary explanations for human behaviours.

Is there progress in evolution? The answer is a qualified 'yes'. More complex types of animals and plants have all evolved from less complex forms, and the history of life shows a general progression from the simplest type of prokaryote single-celled organism to birds and mammals. But there is nothing in the theory of evolution by natural selection to suggest that this is inevitable, and of course bacteria are still one of the most abundant and successful forms of life. This is analogous to the persistence of old, but still useful, tools such as hammers alongside computers in the modern world. In addition, there are many examples of evolutionary reduction of complexity, such as cave-dwelling species that have lost their sight, or parasites that lack the structures and functions needed for independent existence. As we have emphasized several times already, natural selection cannot foresee the future, and merely accumulates variants that are favourable under prevailing conditions. Increased complexity may often provide better functioning, as in the case of eyes, and will then be selected for. If the function is no longer relevant to fitness, it is not surprising that the structure concerned will degenerate.

Evolution is also pitiless. Selection acts to hone the hunting skills and weapons of predators, without regard to the feelings of their prey. It causes parasites to evolve ingenious devices to invade their hosts, even if this causes intense suffering. It causes the evolution of ageing. Natural selection can even cause a species to evolve such a low fertility that it becomes extinct when the environment takes a turn for the worse. Nevertheless, the vision of the history of life revealed in the fossil record, and in the incredible diversity of species alive today, gives a sense of wonder at the results of more than 3 billion years of evolution, despite the fact that this has all resulted 'from the war of nature, from famine and death', in Darwin's phrase. An understanding of evolution can teach us our true place in nature, as part of the immense array of living forms which the impersonal forces of evolution have produced. These evolutionary forces have given our own species the unique ability to reason, so that we can use our foresight to ameliorate the 'war of nature'. We should admire what evolution has produced, and take care not to destroy it through our greed and stupidity, but to preserve it for our descendants. If we fail to do this, our own species could become extinct, along with many other wonderful living creatures.

Further reading

It is well worth reading *On the Origin of Species* by Charles Darwin (John Murray, 1859); the masterly synthesis of innumerable facts on natural history to support the theory of evolution by natural selection is dazzling, and much of what Darwin has to say is still highly relevant. There are many reprints of this available; Harvard University Press have a facsimile of the first (1859) edition, which we used for our quotations.

Jonathan Howard, *Darwin: A Very Short Introduction* (Oxford University Press, 2001) provides an excellent brief survey of Darwin's life and work.

For an excellent discussion of how natural selection can produce the evolution of complex adaptations, see *The Blind Watchmaker: Why The Evidence of Evolution Reveals a Universe without Design* by Richard Dawkins (W.W. Norton, 1996).

The Selfish Gene by Richard Dawkins (Oxford University Press, 1990) is a lively account of how modern ideas on natural selection account for a variety of features of living organisms, especially their behaviour.

Nature's Robots. A History of Proteins by Charles Tanford and Jacqueline Reynolds (Oxford University Press, 2001) is a lucid history of discoveries concerning the nature and functions of proteins, culminating in the deciphering of the genetic code.

Enrico Coen, *The Art of Genes. How Organisms Make Themselves* (Oxford University Press, 1999) provides an excellent account of how genes control development, with some discussion of evolution.

For an account of the application of evolutionary principles to the study of animal behaviour, see *Survival Strategies* by R. Gadagkar (Harvard University Press, 2001).

Richard Leakey and Roger Lewin, *Origins Reconsidered: In Search of What Makes Us Human* (Time Warner Books, 1993) gives an account of human evolution for the general reader.

J. Weiner, *The Beak of the Finch* (Knopf, 1995) is an excellent account of how work on Darwin's finches has illuminated evolutionary biology.

B. Hölldobler and E. O. Wilson, *Journey to the Ants. A Story of Scientific Exploration* (Harvard University Press, 1994) is a fascinating account of the natural history of ants, and the evolutionary principles guiding the evolution of their diverse forms of social organization.

For a discussion of the fossil evidence for the early evolution of life, and experiments and ideas on the origin of life, *Cradle of Life. The Discovery of Earth's Early Fossils* by J. William Schopf (Princeton University Press, 1999) is recommended.

The Crucible of Creation by Simon Conway Morris (Oxford University Press, 1998), which is beautifully illustrated, provides an account of the fossil evidence on the emergence of the major groups of animals.

More advanced books (these assume an A-level knowledge of biology)

Evolutionary Biology by D. J. Futuyma (Sinauer Associates, 1998) is a detailed and authoritative undergraduate textbook on all aspects of evolution.

And a somewhat less detailed undergraduate textbook of evolutionary biology: *Evolution* by Mark Ridley (Blackwell Science, 1996).

Evolutionary Genetics by John Maynard Smith (Oxford University Press, 1998) is an unusually well-written text on how the principles of genetics can be used to understand evolution.

For a comprehensive account of the interpretation of animal behaviour in terms of natural selection, refer to *Behavioural Ecology* by J. R. Krebs and N. B. Davies (Blackwell Science, 1993).

Index

A

B

E

F

G

H

I

J

K

L

O

P

R

S

Expand your collection of

VERY SHORT INTRODUCTIONS

DARWIN

A Very Short Introduction

Jonathan Howard

Darwin's theory of evolution, which implied that our ancestors were apes, caused a furore in the scientific world and beyond when *The Origin of Species* was published in 1859. Arguments still rage about the implications of his evolutionary theory, and scepticism about the value of Darwin's contribution to knowledge is widespread. In this analysis of Darwin's major insights and arguments, Jonathan Howard reasserts the importance of Darwin's work for the development of modern science and culture.

> 'Jonathan Howard has produced an intellectual *tour de force*, a classic in the genre of popular scientific exposition which will still be read in fifty years' time'
>
> ***Times Literary Supplement***

www.oup.co.uk/isbn/0-19-285454-2

ANIMAL RIGHTS

A Very Short Introduction

David DeGrazia

Do animals have moral rights? If so, what does this mean? What sorts of mental lives do animals have, and how should we understand their welfare? After putting forward answers to these questions, David DeGrazia explores the implications for how we treat animals in connection with our diet, zoos, and research.

> 'This is an ideal introduction to the topic. David DeGrazia does a superb job of bringing all the key issues together in a balanced way, in a volume that is both short and very readable.'
>
> **Peter Singer, Princeton University**

> 'Historically aware, philosophically sensitive, and with many well-chosen examples, this book would be hard to beat as a philosophical introduction to animal rights.'
>
> **Roger Crisp, Oxford University**

www.oup.co.uk/isbn/0-19-285360-0

COSMOLOGY

A Very Short Introduction

Peter Coles

What happened in the Big Bang? How did galaxies form? Is the universe accelerating? What is 'dark matter'? What caused the ripples in the cosmic microwave background?

These are just some of the questions today's cosmologists are trying to answer. This book is an accesible and non-technical introduction to the history of cosmology and the latest developments in the field. It is the ideal starting point for anyone curious about the universe and how it began.

> 'A delightful and accesible introduction to modern cosmology'
>
> **Professor J. Silk, Oxford University**

> 'a fast track through the history of our endlessly fascinating Universe, from then to now'
>
> **J. D. Barrow, Cambridge University**

www.oup.co.uk/isbn/0-19-285416-X

DRUGS

A Very Short Introduction

Leslie Iverson

The twentieth century saw a remarkable upsurge of research on drugs, with major advances in the treatment of bacterial and viral infections, heart disease, stomach ulcers, cancer, and mental illnesses. These, along with the introduction of the oral contraceptive, have altered all of our lives. There has also been an increase in the recreational use and abuse of drugs in the Western world. This book explains what drugs are, how they work, and how medicines are developed and tested. It also discusses current ideas about why some drugs are addictive, and whether drug laws need reform.

> 'extremely interesting and capable . . . although called a very short introduction, it contains a wealth of information for the interested layman and is exemplary in its accuracy.'
>
> **Malcolm Lader, King's College, London**

> 'a slim but assured and wise volume on drugs. [It] takes up many controversial positions . . . with an air of authority that commands respect. It is difficult to think of a better overview of the field for anyone new to it.'
>
> **David Healy, University of Wales College of Medicine**

www.oup.co.uk/isbn.0-19-285431-3